Pro DAX with Power BI

Business Intelligence with PowerPivot and SQL Server Analysis Services Tabular

Philip Seamark
Thomas Martens

Apress®

Pro DAX with Power BI: Business Intelligence with PowerPivot and SQL Server Analysis Services Tabular

Philip Seamark
UPPER HUTT, New Zealand

Thomas Martens
Hamburg, Hamburg, Germany

ISBN-13 (pbk): 978-1-4842-4896-6
https://doi.org/10.1007/978-1-4842-4897-3

ISBN-13 (electronic): 978-1-4842-4897-3

Managing Director, Apress Media LLC: Welmoed Spahr
Acquisitions Editor: Joan Murray
Development Editor: Laura Berendson
Coordinating Editor: Jill Balzano

Distributed to the book trade worldwide by Springer Science+Business Media New York, 233 Spring Street, 6th Floor, New York, NY 10013. Phone 1-800-SPRINGER, fax (201) 348-4505, e-mail orders-ny@springer-sbm.com, or visit www.springeronline.com. Apress Media, LLC is a California LLC and the sole member (owner) is Springer Science + Business Media Finance Inc (SSBM Finance Inc). SSBM Finance Inc is a **Delaware** corporation.

For information on translations, please e-mail rights@apress.com, or visit http://www.apress.com/rights-permissions.

Apress titles may be purchased in bulk for academic, corporate, or promotional use. eBook versions and licenses are also available for most titles. For more information, reference our Print and eBook Bulk Sales web page at http://www.apress.com/bulk-sales.

Any source code or other supplementary material referenced by the author in this book is available to readers on GitHub via the book's product page, located at www.apress.com/9781484248966. For more detailed information, please visit http://www.apress.com/source-code.

Printed on acid-free paper

To William
—Philip Seamark

To #thewomanilove
—Thomas Martens

Table of Contents

About the Authors

Philip Seamark is an experienced data warehouse (DW) and business intelligence (BI) consultant with a deep understanding of the Microsoft stack and extensive knowledge of DW methodologies and enterprise data modeling. Recognized for his analytical, conceptual, and problem-solving abilities, he has more than 25 years of commercial experience delivering business applications across a broad range of technologies. His expertise runs the gamut from project management, dimensional modeling, performance tuning, ETL design, development and optimization, and report and dashboard design to installation and administration.

In 2017, Philip received a Microsoft Data Platform MVP award for his contributions to the Power BI community site, and in 2018 he was named a top 10 influencer in the Power BI space. He can be found speaking at many data, analytic, and reporting events around the world. He is also the founder and organizer of the Wellington Power BI User Group. He can be reached through Twitter @PhilSeamark.

Thomas Martens has more than 20 years of experience in the fields of BI, DW, and analytics. In his current role as business intelligence and analytics principal consultant, he helps large enterprises to implement sustainable analytical applications. In 2018, Thomas received a Microsoft Data Platform MVP award for his contributions to the Power BI community site for the first time. He specializes in the visualization of data and the application of analytical methods to large amounts of data. He recognizes DAX as a powerful query language available in many products and wants to leverage his hard-earned experience in creating solutions to bring DAX users to the next level. He can be reached through Twitter @tommartens68.

PART I

The Foundation

CHAPTER 1

DAX Mechanics

The idea behind this chapter is quite simple. Throughout the last years, we have been asked a lot of questions on how to calculate measures and Calculated Columns. And we also have been asked the following question numerous times: "Why does this not work?"

As soon as we started looking closer to the underlying business problem, we started to often wonder, "Why did they ask this question? It's so simple." Then we realized that things we consider simple are often not that simple for other people.

Providing answers to DAX-related questions on the Power BI forum (`https://community.powerbi.com`) earned me thankful remarks like "You are a legend!" or "Wow, you are a DAX ninja!"

Sure, some of these questions have been challenging to answer, but I'm for sure not a legend and also not a ninja. I do not own and have never owned a black pajama that makes me disappear whenever I want or helps me to hover over the rooftops of the buildings in my hometown. From many conversations with clients and friends of mine, I know that there are many smart people outside who are facing DAX challenges that are beyond their current skills that make them think: "What do I have to do to learn this dark art?"

If you might think DAX is some kind of dark art and reading this book will help you conquer the world, and learn some spells, you'll get disappointed in some way. Yes, you will read some DAX code that will help you conquer the world; at least it will help you create revealing DAX statements, that will help you to discover the full potential that is hidden in your data. But you will certainly not find any spell or curse.

DAX helps tremendously to extract insights from your data. Sometimes this extraction is quite easy, and using one DAX function is sufficient. Unfortunately, there are also those moments where this extraction has to happen forcefully, meaning more than one DAX function is used and the DAX code to create a "simple" measure spans across multiple lines, even multiple pages. It's these DAX statements that may lead to the impression that DAX is an art or some kind of an ancient powerful language spoken by witches, sorcerers, or other mystical folk, but be assured it's not.

© Philip Seamark, Thomas Martens 2019
P. Seamark and T. Martens, *Pro DAX with Power BI*, https://doi.org/10.1007/978-1-4842-4897-3_1

But if DAX is not a "conjuring" language, why are there so many questions out there on all the forums like Power BI and even on Stack Overflow?

From our point of view, most of the time it's the oversight of the simple things, the moving parts as we call them. People are often asking if they overlook some hidden "context transition" or if they have to consider the "shadow filter" more closely. And they are also talking about mastering the evaluation context as if mastering it will be rewarded with a black belt.

The problem here is sometimes there is a hidden context transition and sometimes the shadow filter plays its role. But most of the times, they forget that mastering the "five-finger death punch" means that first they have to understand and master the parts that are the foundation of DAX. These basic parts form the foundation upon which DAX unfolds its "magic."

Why DAX mechanics

As already mentioned, using DAX to solve analytical questions is not a secret science practiced by some initiated few. Instead, we consider it a craft. This is because we think that everyone who is willing to spend time and does not fear some setbacks is able to master this craft. We think it's legit to compare the writing of a DAX statement with the creation of some pottery. As you might know, it can take some time to successfully finish an apprenticeship, and it can even take a lifetime to become a master of a craft (thinking of some Japanese pottery). Sometimes the intricate workings of a complex DAX statement may remind us even more of a Swiss-made masterpiece measuring valuable time than a simple mug of coffee that helps us keep awake while we figure out how the moving parts are linked together by our DAX statement.

But nevertheless, the pieces that have to work together flawlessly are few and the laws or principles that rule these pieces – the moving parts – are not that complex or difficult as the rules that command the movement of the planets. For this reason, we find it reasonable to try to demystify the writing of DAX and compare it with a craft that can be mastered.

The moving parts

Before we can start to write DAX, we need an environment that is able to execute a DAX statement and that also provides us with an interface where we can write the DAX statement. For the purpose of this book, this environment is set by using Power BI Desktop that can be downloaded at `www.powerbi.com`.

Power BI Desktop comes with a database (that stores the data and is able to understand/execute DAX queries) and also provides the interface to write these DAX statements. In addition to this and maybe the most obvious part of Power BI Desktop are the many ways to visualize the data stored in the database. Throughout this book, we are referring to a DAX statement or a DAX query, and most of the time it doesn't matter. But there are subtle differences:

- DAX statement

  The term DAX statement is used whenever we refer to a piece of DAX code that is used to define a measure or a Calculated Column.

- DAX query

  The term DAX query is used whenever we are referring to a query that is "automatically" created/composed by Power BI Desktop to retrieve the data from the database to "populate" a visual, no matter if it's a card visual, a table visual, or a clustered column chart.

The database

Much can be said about the database that helps us to find answers to critical questions from various departments throughout the organization, no matter if these organizations are large enterprises or small companies.

This database is Analysis Services Tabular, and its engine is officially called "*xVelocity in-memory analytics engine.*" This engine provides two modes for accessing data:

- Local data

  The data is stored inside the database in an in-memory columnar data store; this mode is commonly known as VertiPaq.

- Remote data

    The data is queried from the data source; this mode is commonly known as DirectQuery .

For the sake of simplicity here, it is called either VertiPaq or even simpler just Analysis Services *Tabular*; this is derived from the term "Business Intelligence Semantic Model Tabular," a name introduced with the release of SQL Server 2012 to differentiate the two analytical engines that have been available since then with SQL Server.

VertiPaq provides its power to the following products inside the Microsoft Business Intelligence offering:

- MSFT SQL Server Analysis Services (SSAS; on premises, since SQL Server 2012)

- Azure Analysis Services

- Power BI Desktop

- Power BI Service

- Power Pivot (in combination with MSFT Excel 2010 until MSFT Excel 2016)

When not explicitly mentioned, we always refer to the version that comes with Power BI Desktop. This book is not meant to cover all the technical details of the VertiPaq engine. This by itself would cover another book, but two points have to be mentioned:

- The data is stored in a *columnar* structure.

- The data is kept in memory.

The columnar and in-memory storage of the data sets the VertiPaq engine apart from SQL Server Relational and from SQL Server Analysis Services Multidimensional (MD). One might think that the in-memory storage limits the size of the dataset that can be stored and analyzed. But in comparison with the row-based data storage of relational database engines, it is possible to compress the data by magnitudes.

But nevertheless, here we will focus on the objects that are more obvious to you, the Power BI user. These objects are

- Tables

- Relationships between these tables

- Measures and Calculated Columns

We use the following picture (Figure 1-1) of a schematic table to explain the workings of certain DAX statements.

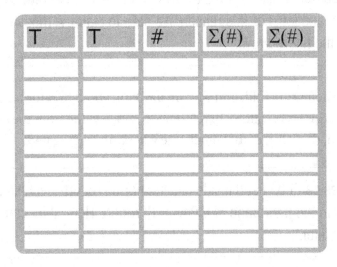

Figure 1-1. *Schematic table*

The preceding table has five columns:

- Two Text Columns (T) – These columns are used to describe the data like *customer name* or *product name.*

- One Numeric Column (#) – This should not be aggregated. This also applies to the column of the data types *datetime* and *date*. These columns often represent key values inside the source system like order *numbers.*

- Two Numeric Columns ($\Sigma(\#)$) – These columns represent columns of numeric data types, like integers or decimal values. These columns will be aggregated and are most often used in measures.

Relationships are essential for the data analysis and play a vital part for performant DAX statements and will be treated extensively in Chapter 2, "Data Modeling." Basically, they relate the tables within a Tabular data model.

Measures are maybe the most powerful feature inside the xVelocity engine. This is simply due to the fact that whenever the data inside the table is not sufficient to extract the insight that we need, we are using DAX to create a calculation.

Definition   A measure returns a scalar value; this means a single value. This scalar value is computed based on the rows of a table, which remain after all the filters have been applied. For this reason, it's safe to claim: A measure is computed by aggregating the filtered rows.

If you find the definition odd, and you are thinking about iterator functions like SUMX, where the expression allows to reference a single value from the current row inside the iteration, don't forget that finally the values are aggregated.

Note   Measures can't be used inside slicers, nor as report-level filter and page-level filter in the filter pane of a Power BI report. Here, they can just be used as a Visual level filter.

Calculated Columns add additional analytical power to the table. Using Power BI Desktop to create the data model, one always has to answer the question if an additional column should be created using Power Query or DAX. Sometimes it seems simpler to create the columns using DAX, but there is a price that has to be paid whenever DAX is used for column creation.

- Calculated Columns created by using DAX will extend the duration needed to process the model.

- Calculated Columns created by using DAX will not compress as good as columns created by Power Query.

The preceding text is not a general recommendation for not using DAX to add columns to the data model. There may be situations where adding columns using DAX reduces the overall time spent on data refresh and model processing until the model is ready to be used for analysis.

One has to be aware of the fact that each column adds to the memory footprint of the data model. For this reason, you might consider to create measures in the future whenever possible.

> **Note**   Calculated Columns can be used on slicers, as report-level filter, and also page-level filters in Power BI reports, and form the content of the categorical axis of the visuals in Power BI.

Power BI Desktop

One of the greatest features of the Analysis Services database is the possibility to add Calculated Columns and measures to the data model. Besides this interface to the database engine, Power BI creates the stage that lets our DAX statements shine, the visuals. These visuals come with their own twist. They provide row and column headers as the Matrix visual does, or an x-axis for categorical values (everything besides fields with a numerical datatype). Sometimes it will become as difficult to show the data that we want to be visualized as it has been to create the measure itself. Chapter 8, "Using DAX to Solve Advanced Reporting Requirements," combines data modeling techniques with DAX statements to create a visual that

- Shows the last N-months in a clustered column chart and the user has to be able to select a certain month that will be used as anchor

- Shows next to the columns of the Top N-customers one additional column that represents the value of all other columns

For now it's sufficient to always remember the "level of interaction" of the objects that we create using DAX, namely, *Calculated Columns* and *measures. What this means is described in* Figure 1-2.

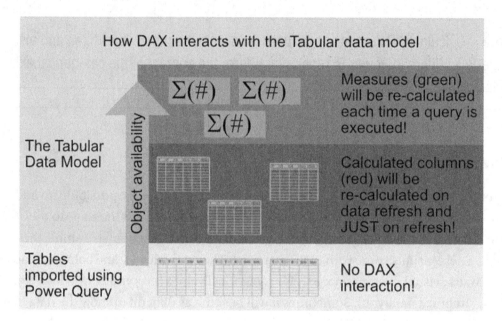

Figure 1-2. *How DAX interacts with the Tabular data model*

What can be learned from Figure 1-2 is the following:

Rule   Objects created in the Tabular data model using DAX are not available from Power Query.

Calculated Columns will not be recalculated if a query is executed. The result of the DAX statement will persist in the underlying table if the DAX statement is initially committed and during data refresh.

Calculated tables created outside of a measure definition will be "created" inside the data model, meaning the DAX statement is executed during the initial creation or whenever the definition changed and of course during data refresh.

DAX: First contact

Throughout the book we will use the data model "Wide World Importers" (WWI) that is available on Git to demonstrate DAX statements with a data model, that is not too simple. The dataset used in this chapter is much easier. I think this is necessary to better understand what is really happening.

Implicit filters

Before we start creating our first DAX statement, it's necessary to have a look at the underlying data of the simple data model used in this chapter. Figure 1-3 shows the content of the table *"simple table values."*

DateRunningIndex	Date	ProductIndex	Product Key	Brand	Color	Amount
384	Monday, March 19, 2018	3	P3	B3	white	13
571	Saturday, September 22, 2018	4	P4	B1	white	13
529	Saturday, August 11, 2018	6	P6	B3	red	14
92	Wednesday, May 31, 2017	6	P6	B3	red	12
357	Tuesday, February 20, 2018	5	P5	B2	blue	11
591	Friday, October 12, 2018	5	P5	B2	blue	7
902	Monday, August 19, 2019	8	P8	B4	blue	6
228	Saturday, October 14, 2017	7	P7	B2	blue	6
378	Tuesday, March 13, 2018	4	P4	B1	white	10
912	Thursday, August 29, 2019	1	P1	B1	red	9
555	Thursday, September 6, 2018	7	P7	B2	blue	12
545	Monday, August 27, 2018	2	P2	B2	blue	9
739	Saturday, March 9, 2019	5	P5	B2	blue	9
546	Tuesday, August 28, 2018	7	P7	B2	blue	11
6	Monday, March 6, 2017	7	P7	B2	blue	7
358	Wednesday, February 21, 2018	2	P2	B2	blue	9
441	Tuesday, May 15, 2018	8	P8	B4	blue	11
160	Monday, August 7, 2017	5	P5	B2	blue	11
707	Tuesday, February 5, 2019	1	P1	B1	red	7
89	Sunday, May 28, 2017	5	P5	B2	blue	8
639	Thursday, November 29, 2018	2	P2	B2	blue	13
725	Saturday, February 23, 2019	8	P8	B4	blue	10
567	Tuesday, September 18, 2018	8	P8	B4	blue	12
848	Wednesday, June 26, 2019	4	P4	B1	white	9

Figure 1-3. *Table – simple table values*

If this table is used in a Matrix visual with the following settings

- Brand column as rows

- Color column as columns

- Amount column as values

you will get what is shown in Figure 1-4 (except the circular marks).

Brand	blue	red	white	Total
B1		16	32	48
B2	113			113
B3		26	13	39
B4	39			39
Total	152	42	45	239

Figure 1-4. *simple table values matrix*

In his book *Beginning DAX with Power BI*, Phil has explained that Power BI adds the values from the row header Brand and from the column header Color to the evaluation context of the measure used on the values band of the Matrix visual.

Note   Filters that are derived from column and row headers or slicer selections are called implicit filters. This is also true for the values that are used on the x-axis of the clustered column chart (this logic can be transferred to all other visuals).

I guess you are not surprised about the value displayed at the intersection of B3/red, even if we did not have defined any measure. The result can be checked easily by just filtering the table in the Data view, see Figure 1-5.

DateRunningIndex	Date	ProductIndex	Product Key	Brand	Color	Amount
92	Wednesday, May 31, 2017	6	P6	B3	red	12
529	Saturday, August 11, 2018	6	P6	B3	red	14

Figure 1-5. *simple table values – B3/red filtered*

It's easy to check that the addition of the values from the column *Amount* equals 26. And we can deduce the following:

- A visual filters the table which contains the column used as value, in this case the column *Amount*.

- An aggregation function is used to compute the value 26 from the two remaining rows after filters have been applied.

It's obvious that the aggregation function is SUM. SUM is the default aggregation function that is applied whenever a numeric column is used as value. The default function can be changed for each column and at least should be checked. In the

Properties ribbon of the Modeling menu, the value for Default Summarization can be changed if necessary. The function can be accessed from the Report or Data view; the column that has been checked has to be marked.

Figure 1-6 shows how the default summarization can be changed for the selected field Amount.

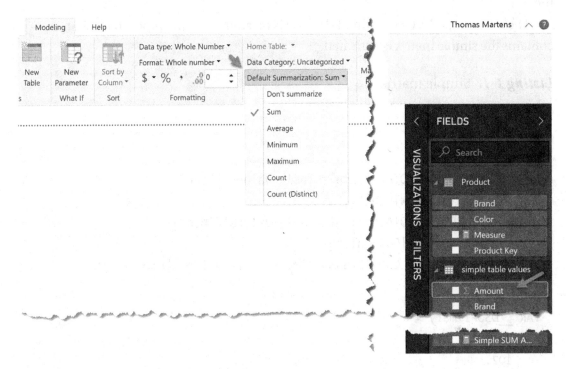

Figure 1-6.   *Column default aggregation function*

Note   If you change the default summarization of a column, you have to remove the column from the visual and add it back.

I have to admit that I have been thinking for quite some time that for each cell inside the Matrix visual including the total values on rows and columns, a separate DAX query is created, because the filter context changes for each cell. Fortunately, this is not the case. What really happens behind the covers can be controlled using DAX Studio, an open source tool that is mainly developed by the guys from sqlbi.com and Darren Gosbell. DAX Studio is an essential tool for the creation of DAX statements (the formatting is much smarter) and whenever you are not satisfied with the performance of your DAX statement. For this reason, some of the capabilities of DAX Studio are described in Chapter 12, "DAX Studio" – at least the functions that are necessary to optimize slow-performing DAX statements.

To discover what's going behind the scenes, DAX Studio is able to catch the DAX query/queries that is/are created by Power BI Desktop, to retrieve the data to populate the visual.

The following DAX (see Listing 1-1) code is created when the report page that contains the simple matrix is activated.

Listing 1-1. Simple matrix

```
DEFINE
  VAR __DSOCore =
    SUMMARIZECOLUMNS(
      ROLLUPADDISSUBTOTAL('simple table values'[Brand],
      "IsGrandTotalRowTotal"),
      ROLLUPADDISSUBTOTAL('simple table values'[Color],
        "IsGrandTotalColumnTotal"),
      "SumAmount", CALCULATE(SUM('simple table values'[Amount]))
    )

  VAR __DSOPrimary =
    TOPN(
      102,
      SUMMARIZE(__DSOCore, 'simple table values'[Brand],
      [IsGrandTotalRowTotal]),
      [IsGrandTotalRowTotal],
      0,
      'simple table values'[Brand],
      1
    )

  VAR __DSOSecondary =
    TOPN(
      102,
      SUMMARIZE(__DSOCore, 'simple table values'[Color],
      [IsGrandTotalColumnTotal]),
```

```
    [IsGrandTotalColumnTotal],
    1,
    'simple table values'[Color],
    1
  )

EVALUATE
  __DSOSecondary

ORDER BY
  [IsGrandTotalColumnTotal], 'simple table values'[Color]

EVALUATE
  NATURALLEFTOUTERJOIN(
    __DSOPrimary,
    SUBSTITUTEWITHINDEX(
      __DSOCore,
      "ColumnIndex",
      __DSOSecondary,
      [IsGrandTotalColumnTotal],
      ASC,
      'simple table values'[Color],
      ASC
    )
  )

ORDER BY
  [IsGrandTotalRowTotal] DESC, 'simple table values'[Brand], [ColumnIndex]
```

We will delve into such DAX queries in much more detail in Chapter 12, "DAX Studio." But for now I just want to direct your attention to the first section of the DAX query. Here will see the following code snippet as part of the DAX function SUMMARIZECOLUMNS.

Listing 1-2. Implicit measure definition

```
"SumAmount", CALCULATE(SUM('simple table values'[Amount]))
```

If we would have written the measure, it would look almost the same. For this reason, we create a measure inside the table *simple table values* using this DAX statement without using the function CALCULATE (see Listing 1-3).

Listing 1-3. Measure – Simple SUM Amount

```
Simple SUM Amount =
SUM('simple table values'[Amount])
```

If you want to check the result, you will realize that we retrieve the same values as from the first query, but this is what we have been expecting.

But what we are looking for is the appearance of our first measure in the query created by Power BI. The following listing shows the first part of the query. And we can see that a "natural" column is treated like an explicitly defined measure:

```
VAR __DS0Core =
    SUMMARIZECOLUMNS(
      ROLLUPADDISSUBTOTAL('simple table values'[Brand],
      "IsGrandTotalRowTotal"),
      ROLLUPADDISSUBTOTAL('simple table values'[Color],
      "IsGrandTotalColumnTotal"),
      "SumAmount", CALCULATE(SUM('simple table values'[Amount])),
      "Simple_SUM_Amount", 'simple table values'[Simple SUM Amount]
    )
```

Until now, you have to believe me that both definitions are the same even if it seems that a CALCULATE function is missing that is wrapped around our measure, unless you have already read Phil's book or some other great DAX books that are available. But we could also prove that both measures yield exactly the same result just by removing the CALCULATE and executing the query in DAX Studio.

Before I will explain what an explicit filter is, I just want to slightly modify the Matrix visual. You will find this matrix on the report page "Chapter 1 – implicit filters – b." In addition to the implicit filters that will be applied to the query execution, I also added the column *Brand* as a page-level filter (just do avoid unwanted interference with other report pages). See Figure 1-7 for the filter settings.

Figure 1-7. *Page-level filter Brand not like B4*

This is an important task because there are some misunderstandings about the behavior of the report-level filter, page-level filter, and also visual-level filter. The following is configured: Remove rows where the value of the column *Brand* equals *B4*. Listing 1-4 shows the query created by Power BI Desktop (at least the important part).

Listing 1-4. FilterTable

```
VAR __DSOFilterTable =
   FILTER(
     KEEPFILTERS(VALUES('simple table values'[Brand])),
     'simple table values'[Brand] <> "B4"
   )

 VAR __DSOCore =
    SUMMARIZECOLUMNS(
      ROLLUPADDISSUBTOTAL('simple table values'[Brand],
      "IsGrandTotalRowTotal"),
      ROLLUPADDISSUBTOTAL('simple table values'[Color],
      "IsGrandTotalColumnTotal"),
```

```
    __DSOFilterTable,
    "SumAmount", CALCULATE(SUM('simple table values'[Amount]))
  )
```

What's important to notice here is the creation of a variable called __*DSOFilterTable* that will be used in all subsequent sections of the query.

To make things much more exciting, I will utilize a third report page "Chapter 1 – implicit filters – c." This report page also has the same page-level filter, but additionally there is also a slicer that is also using the column Brand. Here I will select the Brands B1 and B2. Figure 1-8 shows how the report will look like.

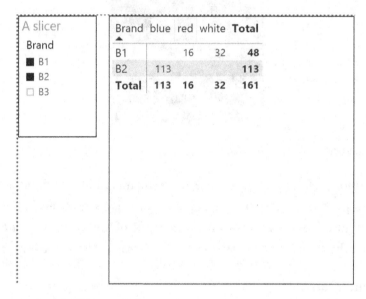

A slicer		Brand	blue	red	white	**Total**
Brand		B1		16	32	**48**
■ B1		B2	113			**113**
■ B2		**Total**	**113**	**16**	**32**	**161**
☐ B3						

Figure 1-8. *Page-level filter and slicer*

Listing 1-5 shows the important parts of the query.

Listing 1-5. FilterTable page-level filter and slicer

```
VAR __DSOFilterTable =
    FILTER(
      KEEPFILTERS(VALUES('simple table values'[Brand])),
      AND(
        'simple table values'[Brand] IN {"B2",
          "B1"},
```

```
      'simple table values'[Brand] <> "B4"
    )
  )
VAR __DSOCore =
  SUMMARIZECOLUMNS(
    ROLLUPADDISSUBTOTAL('simple table values'[Brand],
    "IsGrandTotalRowTotal"),
    ROLLUPADDISSUBTOTAL('simple table values'[Color],
    "IsGrandTotalColumnTotal"),
    __DSOFilterTable,
    "SumAmount", CALCULATE(SUM('simple table values'[Amount]))
  )
```

You may wonder why this gets me so excited or why I find the variable __DS0FilterTable so interesting. But the answer to this is quite simple.

It's not possible to create a measure that shows the SUM of Amount for the three brands. What would be necessary is to remove the filter that has been implicitly added from the slicer but keep the filter that is coming from the page-level filter. To create such a measure, it's necessary to already take some precautions in the data model.

Rule   It's not possible to remove filters coming from the slicers but keep the filter coming from the report-level filter or page-level filter.

Explicit filters

Whenever we are tasked with the writing of a measure, we have to tackle the challenge of the evaluation context. The evaluation context describes the context that is present when the measure (but also Calculated Columns) gets evaluated.

There are two components that determine this context:

- Filter context
- Row context

The filter context can be visualized very easily; just create a Matrix visual, and you can watch the filter context in all its beauty. Row headers and column headers are added to the FilterTable and represent the filter context for the evaluation of the DAX formula in the context of the cell. The row context is sometimes not that obvious. Even the simple creation of a Calculated Column can become burdensome. Whenever CALCULATE is used with more than one parameter, we are going the change the filter context by providing explicit filter.

This is so eminent that we have two dedicated chapters for it: Chapter 4, "This Weird Context Thing," and Chapter 5, "Filtering in DAX."

CHAPTER 2

Data Modeling

Introduction

Even if Power BI is around for quite some time, it seems that data modeling is often neglected, meaning often reports unfold their magic just upon a single table. Sometimes tremendous effort is spent to gather data from various sources to create a single table, and then this single table is used to create appealing visuals.

But then, out of a sudden, it seems to become overly complex to create a measure, or even worse, the measure composed returns an unexpected result.

To fully utilize the analytical potential of DAX, it is necessary to understand some aspects of the Tabular data model at least to some extent. The main goal of this chapter is to introduce important data modeling aspects.

Quite often, a Power BI project has to provide solutions for the following tasks:

- Calculate a year-to-date.
- Calculate the average sales of the last six months to compare it with the current month.

Over time this has led to an impressive collection of patterns that are available on the Internet and also in books. But sometimes these patterns cannot be applied to a particular question because there is a particular twist that has to be considered. The calculation of a measure called "average sales of the last six months" can be challenging as there are not enough data points available, meaning for early data points, there are no "last 6 months." Using only the available data to calculate the average is not sufficient because it will not reflect our business case "the average of 6 months" or does not correspond to our business model. And suddenly, the calculation is facing an additional complexity. It's not just to find the last six months needed for the calculation, but it's also about how to avoid the calculation if there is not enough data available.

© Philip Seamark, Thomas Martens 2019
P. Seamark and T. Martens, *Pro DAX with Power BI*, https://doi.org/10.1007/978-1-4842-4897-3_2

If until now you have been using DAX only to create calculations on top of a single table, this chapter hopefully will provide you with some new ideas and will show why it is almost mandatory to develop datasets that consist of more than one table.

What is a data model

As there are three aspects regarding "thinking in data models," it's necessary to briefly explain quite shortly what a data model is before these aspects will be explained in more detail especially in what this means for creating a data model in Power BI.

- The business process

- The logical data model

- The technical implementation

Many books have been written about data modeling focusing on a different topic or specializing in a particular area. Some books are focusing on high-level data modeling, meaning providing concepts that allow creating a mutual understanding between business people and IT people. Some books are focusing on a particular aspect that is relevant to one database. And some other books are providing techniques on how to transform a business-oriented data model into a physical data model that suits a particular database from a technical point of view.

A data model is a representation of one or more business processes, with the goal to describe how data has to be captured and stored allowing to reflect on the business processes and answer certain questions.

In the analytical realm, there are mainly two different approaches to create a data model; one is called "snowflake schema," and the other one is called "star schema." This book is not the place to argue about one or the other and also not to advocate for one or against the other. A lot of products in the BI world that are close to the business user seem to favor data that is modeled using a star schema. Power BI, Power Pivot, and SQL Server Analysis Services MD and Tabular do not make a difference. For this reason, there is a focus on data modeling following the "star schema" approach.

No matter what approach, what concept you favor, all concepts have at least this one fact in common:

A data model consists of more than one table.

It may seem odd that this simple statement is emphasized, as this seems to be a universal truth, but there are a lot of questions out there, circling around this one theme, how and also why – it is necessary (not so say mandatory) to create a data model.

Star schema

Delving deeper into data modeling for analytical solutions, sooner or later one will discover the concept of the star schema and the underlying technique of dimensional modeling. Dimensional modeling is a technique made popular by Ralph Kimball and his colleagues from the Kimball Group (`www.kimballgroup.com`). Even if the Kimball Group is not active any longer, the content of the web site still provides valuable information.

The name for this kind of data model is derived from the shape the tables of a data model can be arranged into, as in Figure 2-1.

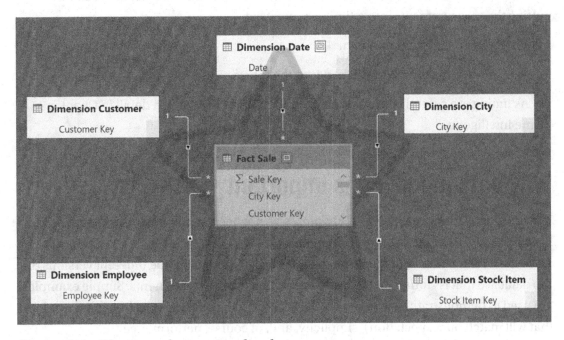

Figure 2-1. *The star schema visualized*

Figure 2-1 shows the data model from the pbix file "CH 2 – relationships – Star Schema.pbix". A data model using dimensional modeling techniques comprises two kinds of tables:

- Dimension tables

- Fact tables

A dimension table represents a particular concept or business object, like customers or products, whereas the fact table contains measurements of a business process in the context of the business objects. This allows analyzing the measurements by using the dimension tables for filtering and slicing and dicing. Dimension tables that are quite common in analytical data models are customer and product tables. But of course this depends on the nature of the business process.

Besides the abovementioned business object, there is another fundamental concept, the concept of time. This concept, or more simply named the calendar table, allows putting the measurements into a timing context like

- Comparing now and then

- Projecting the now and then into the future

As timing is a central concept in data analysis, it is discussed in detail in Chapter 9, "Time Intelligence."

Why data modeling is important

If all the things said in the previous chapter have not been as convincing as they should have been, now it's time to delve deeper into some aspects of the Power BI data model. It's necessary to remember that these details are also valid for all the different versions of the database that is storing the data and evaluating the DAX statements. Simple examples are used to demonstrate how a data model is influencing query results (meaning results that will match our expectation), simplicity, and, of course, performance.

Correct results: Merged filter from a single table

Note   This section is using the pbix file "CH 2 – Auto-Exists.pbix".

One of the great concepts or features of SQL Server Analysis Services Multidimensional (SSAS MD) is the concept of Auto-Exist, but maybe some of the details have become a little faded out. For this reason, here is a recap.

Auto-Exist prevents SSAS Multidimensional from returning nonexisting combinations of attributes if one or more attributes from the same dimension are used in the same query. Translation to SSAS Tabular or Power BI data models goes like this. A query will not return nonexisting combinations of columns (attributes) from tables (dimensions). As there are just tables in a Power BI data model, it's a valid approach to think of tables on the one side of relationship (relationships will be discussed in more detail later on in this chapter) that filter tables on the many side of the relationship.

The important point here is this concept does also exist in Power BI and of course in Power Pivot and SSAS Tabular. It's important to always consider this fact, even if it is not obvious right now.

The following SQL statement (see Listing 2-1) is used to create a view from three different tables of the Wide World Importers DW database.

Listing 2-1. Auto-Exist – SingleTable

```
CREATE VIEW
        [Fact].[v_sales_singletable]
AS
SELECT
        dimDate.[Calendar Year]
,       i.Size
,       COUNT(f.[Sale Key]) AS NoOfSales
FROM [Fact].[Sale] AS f
        INNER JOIN Dimension.Date AS dimDate ON
                f.[Invoice Date Key] = dimDate.Date
        INNER JOIN Dimension.[Stock Item] AS i ON
                f.[Stock Item Key] = i.[Stock Item Key]
GROUP BY
        dimDate.[Calendar Year]
,       i.size
```

The data has been imported to the Power BI file "CH 2 – Auto-Exists.pbix". Figure 2-2 shows the columns of the table.

Figure 2-2. *The single table*

This table represents sales orders (NoOfSales) of various product sizes (Size) for different calendar years (Calendar Year). This table is just to prove the point that under certain circumstances, a DAX statement will not return the expected result. Even if this table just reflects a random combination of two columns from different business objects (product and time), it is more than likely that this will happen sooner or later in real life. Unfortunately, the hidden issue will not surface immediately, but as soon as more complex questions have to be answered on a larger dataset, the unexpected result, or the wrong result, may lead to wrong decisions.

The next screenshot shows a little portion from the report page "Ch 02 – Auto Exist and a Single Table."

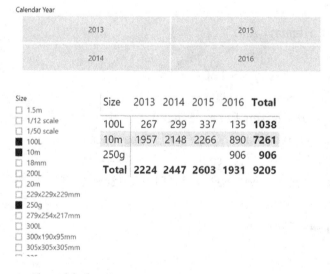

Figure 2-3. *The single table basic report*

Assuming that the following measurements have to be defined to answer some business questions

- One measure that counts the sizes in the current selection
- One measure that counts the sizes disrespecting the selection of the year

a sample visualization of these measurements is shown in Figure 2-4.

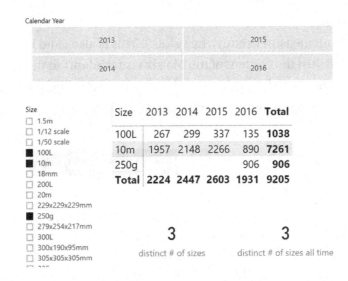

Figure 2-4. *The single table basic report – measures*

Listings 2-2 and 2-3 are showing the DAX statements used to define the measures.

Listing 2-2. Auto-Exist – SingleTable, distinct # of sizes

```
distinct # of sizes =
DISTINCTCOUNT('Fact v_sales_singletable'[Size])
```

Listing 2-3. Auto-Exist – SingleTable, ms 2, distinct # of sizes all time

```
distinct # of sizes all time =
CALCULATE(
    [distinct # of sizes]
    ,ALL('Fact v_sales_singletable'[Calendar Year])
)
```

As one can see from Listing 2-3, inside the CALCULATE(...), the ALL(...) function is used to remove any existing filter from the column Calendar Year. Chapter 5, "Filtering in DAX," will explain in great detail all the intricacies of filtering. For this reason, the explanation here is just this: The DAX function CALCULATE(...) allows to alter an existing filter context. This means that the first parameter of the CALCULATE function, the expression, will be evaluated after all the other parameters have been applied to change the existing filter context. One of the DAX functions to alter the existing filter context is ALL. ALL removes an existing filter context either from a whole table or just from a single column.

In Figure 2-4, both measures return the value 3. This is also valid if we select Calendar Year : 2016. Just the content of the Matrix visual adapts to the current selection. But if we choose Calendar Year : 2015 from the slicer, the value for the measure "distinct # of sizes all time" also changes (see Figure 2-5).

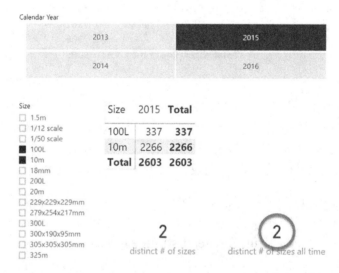

Figure 2-5. *The single table basic report – wrong expectation*

Talking about expectations and from the explanation given on how the measure `distinct # of sizes all time` works, the expected result should be 3, as the measure should return the number of sizes no matter what year has been selected from the slicer.

To understand what's going on, it's necessary to capture the DAX query that is created by Power BI and sent to the data model to populate each visual. For this, DAX Studio is used. DAX Studio is an essential tool not just for optimizing DAX statements, but also to understand what's going on, and to really understand some subtle differences in similar DAX functions. For this reason, Chapter 12, "DAX Studio," discusses how DAX Studio can be used to gain a better understanding of DAX and for optimizing DAX statements.

Note   It's essential to keep in mind that each visual is populated by its own DAX query. The more visuals are used on a single report page, the more DAX queries will be generated.

This said, let's delve right into the DAX statement that is passed when Calendar Year : 2015 is selected from the slicer.

Listing 2-4. Auto-Exist – SingleTable, DAX query sent

```
DEFINE
  VAR __DSOFilterTable =
    TREATAS({2015}, 'Fact v_sales_singletable'[Calendar Year])

  VAR __DSOFilterTable2 =
    TREATAS({"100L",
      "10m",
      "250g"}, 'Fact v_sales_singletable'[Size])
EVALUATE
  SUMMARIZECOLUMNS(
    __DSOFilterTable,
    __DSOFilterTable2,
    "distinct__of_sizes_all_time", IGNORE('Fact v_sales_
    singletable'[distinct # of sizes all time])
  )
```

If this DAX statement is executed inside DAX Studio, the query returns 2.

Why is this?

The answer is not simple to understand but is due to the Auto-Exist concept that is considered by the function SUMMARIZECOLUMNS(...). If two columns from the same table are used, just existing combinations of the values inside the table are returned. Due to the fact that the combination of [Calendar Year] = 2015 and [Size] = 250 g does not exist in the table, only the values 100 L and 10 m are considered by the measure that counts the distinct occurrences of the size, and for this reason, the query returns 2.

To check the rows before the measure is applied, it's necessary to rewrite the DAX query from Listing 2-4. This query will look like Listing 2-5.

Listing 2-5. Auto-Exist – SingleTable, DAX rewritten

```
EVALUATE
CALCULATETABLE(
  SUMMARIZE('Fact v_sales_singletable','Fact v_sales_singletable'[Calendar
  Year],'Fact v_sales_singletable'[Size])
  ,TREATAS({2015}, 'Fact v_sales_singletable'[Calendar Year])
  ,TREATAS({"100L","10m","250g"}, 'Fact v_sales_singletable'[Size])
)
```

The query from Listing 2-5 returns a table with just two rows, even if the expectation might be different, namely, three, given the ALL(...) function that is used inside the measure (see Figure 2-6).

Results

Calendar Year	Size
2015	100L
2015	10m

Figure 2-6. *The single table basic report – rows considered*

Caution   SUMMARIZE and SUMMARIZECOLUMNS will return only existing combinations if the columns are coming from the same table.

We might think that the ALL(...) should remove the filter that is applied by the slicer selection, but we have to consider that the table already does not contain the Calendar Year 2016 and the size 250 g. This is because the nonexisting combination has already been removed by SUMMARIZECOLUMNS. ALL will remove the existing filter of the column Calendar Year, but it will not bring back the third size 250 g, as this size does not exist in the year 2015.

This behavior is by no means a bug, this is how it is implemented, and in almost each case this is the behavior we want. This implementation is optimized for performance. Only data models that consist of a single table might encounter this problem.

This problem can be avoided if we try to create a proper star schema right from the beginning. For this reason, the pbix file also contains a simple star schema (see Figure 2-7).

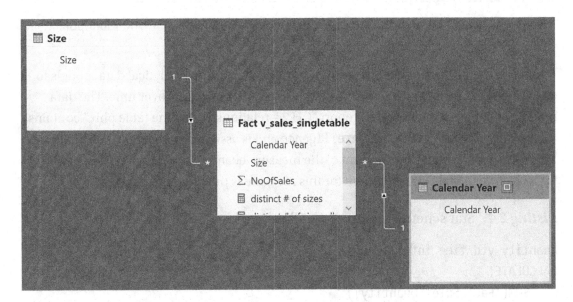

Figure 2-7. *Simple star schema*

To properly use this star schema, we can create a measure as in Listing 2-6.

Listing 2-6. Auto-Exist – star schema, measure rewritten

```
star distinct # of sizes all time =
CALCULATE(
    [distinct # of sizes]
    ,ALL('Calendar Year'[Calendar Year])
)
```

Instead of using the column *Calendar Year* from the table Fact v_sales_ singletable, now this measure is using the ALL(…) in combination with the column from the *dimension* table Calendar Year.

If you take a closer look, the report page "CH 02 – Star Schema" is using the columns from both *dimension* tables *Calendar Year* and *Size*.

A data model that is built using dimensional modeling techniques for this reason comes with another great benefit, and it's just simplicity.

Simplicity: About relationships and filter propagation

Note   This section is using the pbix file "CH 2 – relationships – One table.pbix".

As already mentioned, one of the most important concepts in analytical data models to answer ever-recurring questions is the analysis of measurements over time. The data model contained in the Power BI report "CH 2 – relationships – One table.pbix" contains a measure quantity ytd time intelligence and is assigned to the table "Fact Sale." This measure calculates the amount of the measure quantity until the end of a period for a given year. The DAX statement for this measure is provided in Listing 2-7.

Listing 2-7. Star schema – ytd time intelligence

```
quantity ytd time intelligence =
CALCULATE(
    SUM('Fact Sale'[Quantity])
    ,DATESYTD('Dimension Date'[Date])
)
```

This measure uses the time intelligence function DATESYTD.

The measure "quantity ytd" is calculating the same, but without using a time intelligence function. The DAX for this measure is provided in Listing 2-8.

Listing 2-8. Star schema – ytd base functions

```
quantity ytd =
CALCULATE(
    SUM('Fact Sale'[Quantity])
    ,FILTER(
        ALL('Dimension Date'[Date])
        ,'Dimension Date'[Date] <= MAX('Dimension Date'[Date])
        && YEAR('Dimension Date'[Date]) = YEAR(MAX('Dimension Date'[Date]))
    )
)
```

Even if these DAX statements do not reveal anything new, it's nevertheless worthwhile to mention that this measure can be used in different situations, and it's not necessary to adjust the DAX statement.

Figure 2-8 makes this much clearer.

Cumulative quantity by months

Calendar Month Number	Quantity	quantity ytd	quantity ytd time intelligence
1	193,271	193,271	193,271
2	142,120	335,391	335,391
3	207,486	542,877	542,877
4	212,995	755,872	755,872
5	230,725	986,597	986,597
6	213,468	1,200,065	1,200,065
7	232,599	1,432,664	1,432,664
8	192,199	1,624,863	1,624,863
9	190,567	1,815,430	1,815,430
10	198,476	2,013,906	2,013,906
11	194,290	2,208,196	2,208,196
12	193,461	2,401,657	2,401,657
Total	2,401,657	2,401,657	2,401,657

Cumulative quantity by invoice dates

Date	Quantity	quantity ytd	quantity ytd time intelligence
Friday, February 1, 2013	9,934	203,205	203,205
Saturday, February 2, 2013	4,551	207,756	207,756
Sunday, February 3, 2013		207,756	207,756
Monday, February 4, 2013	8,025	215,781	215,781
Tuesday, February 5, 2013	4,543	220,324	220,324
Wednesday, February 6, 2013	4,523	224,847	224,847
Thursday, February 7, 2013	4,702	229,549	229,549
Friday, February 8, 2013	7,434	236,983	236,983
Saturday, February 9, 2013	4,896	241,879	241,879
Sunday, February 10, 2013		241,879	241,879
Monday, February 11, 2013	8,602	250,481	250,481
Tuesday, February 12, 2013	5,257	255,738	255,738
Wednesday, February 13, 2013	4,867	260,605	260,605
Thursday, February 14, 2013	6,910	267,515	267,515
Total	142,120	335,391	335,391

Figure 2-8. *Star schema – time intelligence working (report page months)*

Note   The following is referring to the Power BI file "CH 2 – relationships – One table.pbix".

Trying to apply a similar logic (applying time intelligence to answer questions) to the data model, we will realize that we cannot use any time intelligence functions. Listing 2-9 shows the adjusted DAX statement for the "one-table solution."

Listing 2-9. One table – ytd time intelligence

```
quantity ytd time intelligence =
CALCULATE(
    SUM('Fact v_starschema_asonetable'[Quantity])
    ,DATESYTD('Fact v_starschema_asonetable'[Date])
)
```

Figure 2-9 shows this does not work. The cumulative value for the second month matches the noncumulative value for the second month.

Calendar Month Number	Quantity	qunatity ytd time intelligence
1	193271	193271
2	142120	142120
3	207486	207486
4	212995	212995
5	230725	230725
6	213468	213468
7	232599	232599
8	192199	192199
9	190567	190567
10	198476	198476
11	194290	194290
12	193461	193461
Total	2401657	2401657

Figure 2-9. *One table – time intelligence not working*

This is due to the fact that the Date column is not used on rows (more details on how filtering really works will be revealed in Chapter 5, "Filtering in DAX"). The DAX statement that returns cumulative values for the month numbers is shown in Listing 2-10.

Listing 2-10. One table – ytd base functions month number

```
quantity ytd =
CALCULATE(
    SUM('Fact v_starschema_asonetable'[Quantity])
    ,FILTER(
        ALL(
                'Fact v_starschema_asonetable'[Calendar Month Number]
                ,'
            )
        ,'Fact v_starschema_asonetable'[Calendar Month Number] <= MAX('Fact
        v_starschema_asonetable'[Calendar Month Number])
    )
)
```

This measure creates the expected values as shown in Figure 2-10.

Calendar Month Number	Quantity	qunatity ytd time intelligence	quantity ytd
1	193,271	193,271	193,271
2	142,120	142,120	335,391
3	207,486	207,486	542,877
4	212,995	212,995	755,872
5	230,725	230,725	986,597
6	213,468	213,468	1,200,065
7	232,599	232,599	1,432,664
8	192,199	192,199	1,624,863
9	190,567	190,567	1,815,430
10	198,476	198,476	2,013,906
11	194,290	194,290	2,208,196
12	193,461	193,461	2,401,657
Total	**2,401,657**	**2,401,657**	**2,401,657**

Figure 2-10. *One table – cumulative month number*

Even if the measure "quantity ytd" returns the correct and expected results, this measure is in many ways inferior to the measure of the star schema. As soon as it becomes necessary to use the column `Calendar Month Label` instead of the column `Calendar Month Number` as row header, the current measure stops working. This is shown in Figure 2-11.

Calendar Month Label	Quantity	qunatity ytd time intelligence	quantity ytd
CY2013-Apr	212,995	212,995	212,995
CY2013-Aug	192,199	192,199	192,199
CY2013-Dec	193,461	193,461	193,461
CY2013-Feb	142,120	142,120	142,120
CY2013-Jan	193,271	193,271	193,271
CY2013-Jul	232,599	232,599	232,599
CY2013-Jun	213,468	213,468	213,468
CY2013-Mar	207,486	207,486	207,486
CY2013-May	230,725	230,725	230,725
CY2013-Nov	194,290	194,290	194,290
CY2013-Oct	198,476	198,476	198,476
CY2013-Sep	190,567	190,567	190,567
Total	**2,401,657**	**2,401,657**	**2,401,657**

Figure 2-11. *One table – cumulative month label*

This also happens if the Date column is used instead. Basically this will happen whatever date-related column is used. It's more or less simple to create a DAX statement that works in the one-table solution no matter which column we want to use as row header in the Matrix visual. This is possible by considering each column in the DAX statement. This is shown in Listing 2-11.

Listing 2-11. One table – ytd base functions a little more final

```
quantity ytd a little more final =
CALCULATE(
    SUM('Fact v_starschema_asonetable'[Quantity])
    ,FILTER(
        ALL(
            'Fact v_starschema_asonetable'[Calendar Month Number]
            ,'Fact v_starschema_asonetable'[Calendar Month Label]
            ,'Fact v_starschema_asonetable'[Date]
        )
        ,'Fact v_starschema_asonetable'[Calendar Month Number] <= MAX('Fact
        v_starschema_asonetable'[Calendar Month Number])
        && 'Fact v_starschema_asonetable'[Date] <= MAX('Fact v_starschema_
        asonetable'[Date])
    )
)
```

To overcome the limitation of a missing table that will be used as a dedicated calendar table, all columns can be inside the ALL function. All columns that are referenced inside the ALL function are also used inside the function FILTER (<table>, <filter>), at least two of the columns.

It's not necessary to consider the column Calendar Month Label also in the filter part, as there is an implicit one-to-one relationship between the columns Calendar Month Label and Calendar Month Number.

The reason behind the growing complexity of measures in a one-table solution in contrast to solutions based on data models that follow the star schema approach is that the latter data models are able to use the concept of relationships and filter propagation. Both of these concepts will be explained in greater detail in the next sections.

Relationships

Note   This section is using the pbix file "CH 2 – relationships – missing relationships.pbix".

As soon as our data is based on more than one table, we have to consider how to relate these tables. The relationships between the tables of our data model can be defined in the "Relationships" view of Power BI Desktop.

In the Power BI file "CH 2 – Star Schema – import from SQL – missing relationships. pbix", two relationships are missing:

- Dimension Stock Item → Fact Sale

- Dimension Date → Fact Sale

The simplest way to define a relationship between two tables is to drag a column from one table to the corresponding column in the other table. The missing relationship between the tables Fact Sale and Dimension Stock Item can be established by dragging the column Stock Item Key from one table to the corresponding column in the other table. It doesn't matter which table will be selected as starting point.

The next way to create a relationship is to use the dialog "Manage Relationships." This dialog can be called from the "Home" menu using the "Manage Relationships" command from the "Relationships" ribbon, as shown in Figure 2-12.

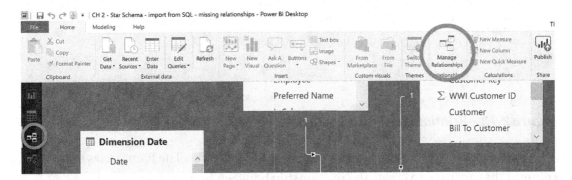

Figure 2-12. *Relationships – Manage Relationships dialog*

Using this dialog, it's possible to select both tables between which a relationship has to be created. Figure 2-13 shows how the relationship between the tables Fact Sale and Dimension Date will be created, using the button "New."

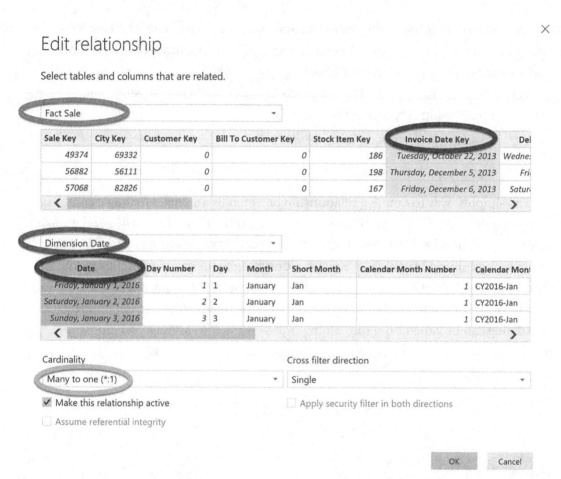

Figure 2-13.   Relationships – using the dialog

The green and violet circles in Figure 2-13 mark the tables. The blue circles mark the columns that are used to create the relationship between both tables.

Note   Just one column can be used from each table to establish the relationship.

The selection "Many to one (∗:1)" in the field Cardinality happened automatically.

Filter propagation

Note   This section is using the pbix file "CH 2 – relationships – Star Schema.pbix".

Filter propagation is an essential concept for the analytical data model, not just for the data model used by the Power BI, but also for the data models in Power Pivot, SQL Server Analysis Services Tabular, and also Azure Analysis Services. For this reason, it's mandatory to understand the possible values of the field "Cardinality" in combination with the values of the property "Cross filter direction" and how these values will impact the general usability and how different values may impact the outcome of measures.

Filter propagation is just simply this. If a table is filtered, for example, by selecting a slicer, the filtered values will be propagated to related tables using the relationship. Selecting the value CY2014 from the column Calendar Year Label of the table Dimension Calendar filters (reduces) the table to all the rows where the value equals the selected value.

By default, this propagation follows the path of the relationship from the one side to the many side of the relationship. Figure 2-14 will help to make this more obvious.

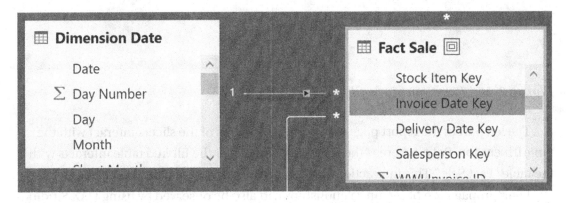

Figure 2-14. *Relationships – one to many*

As the relationship between both tables is established by using the columns 'Dimension Date'[Date] and 'Fact Sale'[Invoice Date Key], the values of the column [Date] that are filtered by any slicer that is based on the Dimension Date table are automatically **propagated** to the table on the many side, in this case to the fact table.

This is one of the ingredients that allow the great performance of slicing and dicing through large amounts of data stored in the Power BI data model (and of course all the other data models).

The Power BI file "CH 2 – relationships – Star Schema.pbix" contains the report page "Filter propagation." This report page has not been created with the intention of creating an awesome-looking report, but with the intention of visualizing filter propagation. Figure 2-15 shows the report page after some values have been selected from the slicers.

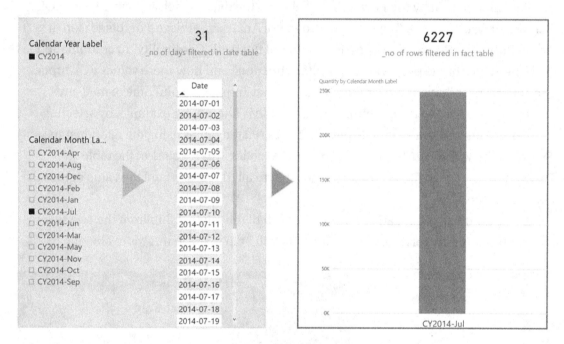

Figure 2-15. *Relationships – filter propagation*

The left part of the report page shows how selections of the slicers interact with the table Dimension Date, whereas the right part shows how the filtered table interacts with the table "Fact Sale." This interaction is based on the concept of filter propagation.

Filter propagation based on relationships can also be observed by using DAX Studio. Listing 2-12 shows the DAX query that retrieves the value for the card visual that depicts the number of rows from the fact table "Fact Sale."

Listing 2-12. Relationships – DAX and xmSQL, DAX

```
DEFINE
  VAR __DSOFilterTable =
    TREATAS({"CY2014"}, 'Dimension Date'[Calendar Year Label])

  VAR __DSOFilterTable2 =
    TREATAS({"CY2014-Jul"}, 'Dimension Date'[Calendar Month Label])

EVALUATE
  SUMMARIZECOLUMNS(
    __DSOFilterTable,
    __DSOFilterTable2,
    "v_no_of_rows_filtered_in_fact_table", IGNORE('Fact Sale'[_no of rows
    filtered in fact table])
  )
```

Even if this has not been mentioned before, and will be covered in more detail in Chapter 12, "DAX Studio," now it's time to mention the two engines that empower the columnar in-memory database. These are the storage and the formula engines. Please don't get tricked by the term "storage engine." This storage engine (SE) retrieves data from RAM and not from disk. There may be situations where tasks that are related to the storage engine also trigger other tasks that then may read data from slower hard drives or faster solid-state drives, but this will not be the case if we are using data that is stored inside our Power BI files.

The formula engine detects what data is needed to answer the question, meaning to provide the result needed to visualize the data and evaluate the complex expressions. The storage engine retrieves the data and passes the data back to the formula engine for further calculation if needed.

The storage engine can be considered as "the muscle" and the formula engine as "the brain." This means that the storage engine can solve its tasks incredibly fast, whereas the formula engine is able to solve complex computations. For this reason, it's always a good idea to use the power of the storage engine whenever possible. DAX optimization basically shifts tasks from the formula engine to the storage engine.

This said, let's consider this as some kind of short introduction. Now we are going back to the preceding DAX query. DAX Studio also provides a feature that captures the activity of the Storage Engine. This activity is expressed as some kind of SQL dialect that is called xmSQL. Listing 2-13 shows this query.

Listing 2-13. Relationships – DAX and xmSQL, xmSQL

```
SET DC_KIND="AUTO";
SELECT
COUNT ( )
FROM 'Fact Sale'
        LEFT OUTER JOIN 'Dimension Date' ON 'Fact Sale'[Invoice Date
        Key]='Dimension Date'[Date]
WHERE
        'Dimension Date'[Calendar Month Label] = 'CY2014-Jul' VAND
        'Dimension Date'[Calendar Year Label] = 'CY2014';
```

The essential takeaway from the xmSQL query shown in Listing 2-13 is the LEFT OUTER JOIN between the columns that have been used to define the relationship.

It can help to "visualize" the concept of filter propagation as passing the values from the column that has been used on the one side of the relationship to the column on the many side of the relationship. This passing of values finally leads to a filtered table. Upon these "remaining" rows, the measure is applied that counts the rows.

Relationship cardinality

The cardinality of relationships plays an important role for the user experience and also how DAX functions unfold their magic and leverage the cardinality of a relationship. This will be explained and demonstrated in the next chapters.

Many to one (*:1)/One to many (1:*)

The relationship between tables with a many side and a one side is the most important relationship type in Power BI and related data models. This relationship type is also called "strong relationship." If you wonder if there are also weak relationships, just read on.

This type of relationship plays an important role for the usability and performance of the data model executing DAX queries and evaluating DAX expressions. For this reason, it's necessary to strive for using this relationship as often as possible.

The existence of relationships and therefore the performance of the data model can be used in our DAX statements. The following listings show some examples of how relationships between tables can be used in DAX.

Listing 2-14. Relationships – DAX, SUMMARIZECOLUMNS

```
EVALUATE
SUMMARIZECOLUMNS(
     'Dimension City'[Sales Territory]
     ,'Dimension Stock Item'[Color]
     ,"No of rows"
     ,COUNTROWS('Fact Sale')
)
```

Figure 2-16 shows a fraction of the returned table when the DAX statement in Listing 2-14 is executed inside SQL Server Management Studio (SSMS).

Dimension City[S...	Dimension Stock...	[No of rows]
Mideast	N/A	14255
Southeast	N/A	21289
Great Lakes	N/A	11023
Plains	N/A	12995
Southwest	N/A	13424
Far West	N/A	11025
Rocky Mountain	N/A	6199
External	N/A	1166
New England	N/A	4434
Mideast	Red	772
Southeast	Red	1126
Great Lakes	Red	621

Figure 2-16. *Relationships – DAX, SUMMARIZECOLUMNS*

Listing 2-15 shows the same query, but now using the older function SUMMARIZE.

Listing 2-15. Relationships – DAX, SUMMARIZE

```
EVALUATE
ADDCOLUMNS(
     SUMMARIZE(
          'Fact Sale'
          ,'Dimension Customer'[Category]
```

```
                ,'Dimension Stock Item'[Color]
    )
    ,"No of rows"
    ,COUNTROWS('Fact Sale')
)
```

SUMMARIZECOLUMNS can be considered as an optimized version of SUMMARIZE that is faster and overcomes some of the limitations of the older SUMMARIZE. Nevertheless, at the moment of this writing, SUMMARIZECOLUMNS cannot be used inside a measure. For this you will find a lot of measures that are still using SUMMARIZE.

Executing both queries using either DAX Studio or SQL Server Management Studio, both queries do not return the same result. This is due to the fact that the function ADDCOLUMNS(...) does not implicitly create a filter context which is used when the expression COUNTROWS(...) gets evaluated. This can be resolved by wrapping a CALCULATE(...) around the COUNTROWS(...).

Further investigation will reveal that SUMMARIZE returns the expected results, even without using ADDCOLUMNS(...), creating the column inside of SUMMARIZE(...). It's absolutely necessary to be aware that under circumstances this will create unexpected results. Using SUMMARIZE as intended has its own difficulties. These difficulties are described in a blog post by the awesome people from sqlbi.com, Alberto Ferrari and Marco Russo: www.sqlbi.com/articles/best-practices-using-summarize-and-addcolumns/.

For this reason, SUMMARIZE should only be used to group columns.

Listings 2-16 and 2-17 show how the DAX functions RELATED and RELATEDTABLE can be used to "pull" values either from the one side to the many side or from the many side to the one side.

Listing 2-16 shows how the RELATED(...) function is used inside a Calculated Column definition inside the table "Fact Sale." The DAX statement pulls the value from the column Unit Price from the table Dimension Stock Item into the table Fact Sale. The syntax is quite easy, as the existing relationship will be leveraged.

One has to keep in mind that RELATED needs an unambiguous row context. For this reason, it can be used either in DAX statements used to create a Calculated Column or inside iterator functions like SUMX(...).

Listing 2-16. Relationships – DAX, RELATED

```
Unit Price 2 =
RELATED('Dimension Stock Item'[Unit Price])
```

Listing 2-17. Relationships – DAX, RELATEDSUMX

```
Sales for 250g items =
SUMX(
    FILTER(
        'Fact Sale'
        ,RELATED('Dimension Stock Item'[Size]) = "250g"
    )
    ,'Fact Sale'[Quantity] * RELATED('Dimension Stock Item'[Unit Price])
)
```

The ...X functions like SUMX(...) or MINX(...) are table iterator functions. These functions are used to iterate across all the rows of a table. Listing 2-17 shows how RELATED(...) can be used in combination with the functions FILTER(...) and SUMX(...). The DAX statement calculates the product of [Quantity] from the table "Fact Sale" and [Unit Price] from the Dimension Stock Item table, but only for items with a size of 250 g.

RELATEDTABLE(...) works similar to RELATED, but with its own twist. RELATEDTABLE(...) returns a table that is transforming an existing row context into a filter context used for the evaluation of the function.

Listing 2-18. Relationships – DAX, RELATEDTABLE

```
No of sales =
COUNTROWS(
    RELATEDTABLE('Fact Sale')
)
```

The result of the Calculated Column from Listing 2-18 is shown in Figure 2-17.

Customer	No of sales
Tailspin Toys (Absecon, NJ)	319
Tailspin Toys (Aceitunas, PR)	344
Tailspin Toys (Airport Drive, MO)	404
Tailspin Toys (Alstead, NH)	315
Tailspin Toys (Amanda Park, WA)	349
Tailspin Toys (Andrix, CO)	356
Tailspin Toys (Annamoriah, WV)	365
Tailspin Toys (Antares, AZ)	369
Tailspin Toys (Antonito, CO)	357
Tailspin Toys (Arbor Vitae, WI)	344
Tailspin Toys (Arietta, NY)	390
Tailspin Toys (Armstrong Creek, WI)	380
Tailspin Toys (Arrow Rock, MO)	377
Tailspin Toys (Ashtabula, OH)	387
Tailspin Toys (Aspen Park, CO)	335
Total	**228265**

Figure 2-17. *Relationships – DAX, RELATEDTABLE result*

How RELATEDTABLE does work can be explained best by comparing the DAX statement from Listing 2-18 with a corresponding SQL statement. This statement is shown in Listing 2-19.

Listing 2-19. SQL statement to reproduce RELATEDTABLE

```
SELECT
        c.[Customer Key]
,       c.Customer
,       COUNT(factsale.[sale key]) AS 'No of Sales'
FROM
        dimension.customer AS c
            LEFT OUTER JOIN fact.Sale AS factsale ON
                    factsale.[Customer Key] = c.[Customer Key]
GROUP BY
        c.[Customer Key]
,       c.Customer
```

RELATEDTABLE utilizes the relationship that exists between the Dimension Customer table and the table Fact Sale. In the SQL statement, the corresponding part is the LEFT OUTER JOIN. As RELATEDTABLE returns a table, it's necessary to create a scalar value as RELATEDTABLE is used to create a Calculated Column. For this reason, RELATEDTABLE is wrapped inside COUNTROWS.

Relationships with the cardinality of "one to many" or "many to one" are the most powerful and important type of relationships.

Rule   Always use 1:∗ (one to many) relationships. If you think it's not possible, think again. If it's still not possible, use a different cardinality.

One to one (1:1)

Note   This section is using the pbix file "CH 2 – relationships – one to one.pbix".

One-to-one relationships are quite uncommon and should not be used, as these relationships provide no analytical functionality.

If tables are imported from relational data sources, it can happen that certain contexts of an analytical object are modeled in different tables. For example, occupation and address can be modeled in two different tables. Using one-to-one relationships prevents using powerful one-to-many relationships. It is strongly recommended to combine both tables into a single table.

Figure 2-18 shows an example of this type of relationship. Please be aware that this is just an example.

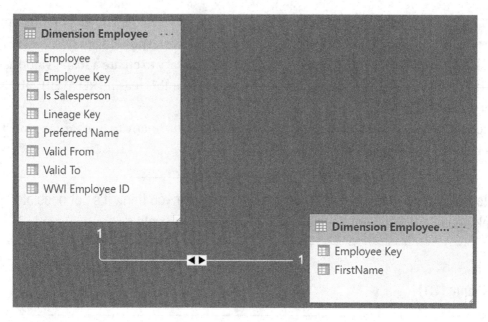

Figure 2-18. *Relationships – one-to-one EmployeeFirstName details*

Many to many (∗:∗)

This relationship type allows to relate tables, even if no column used to create the relationship has unique values.

Note   This section is using the pbix file "CH 2 – relationships – many to many. pbix".

Figure 2-19 shows a simple example using a simple fact table with just three columns and the table Customer Hobby (m2m).

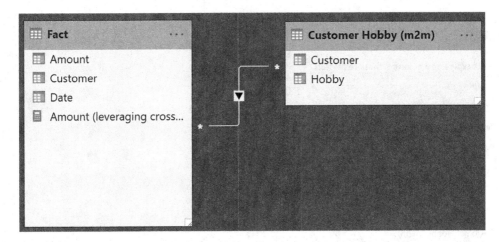

Figure 2-19. *Relationships – many to many*

These tables allow to extract insights between the relationships of customers and their hobbies and the corresponding facts.

Figure 2-20 shows the relationship dialog when the Customer columns are used to create a relationship.

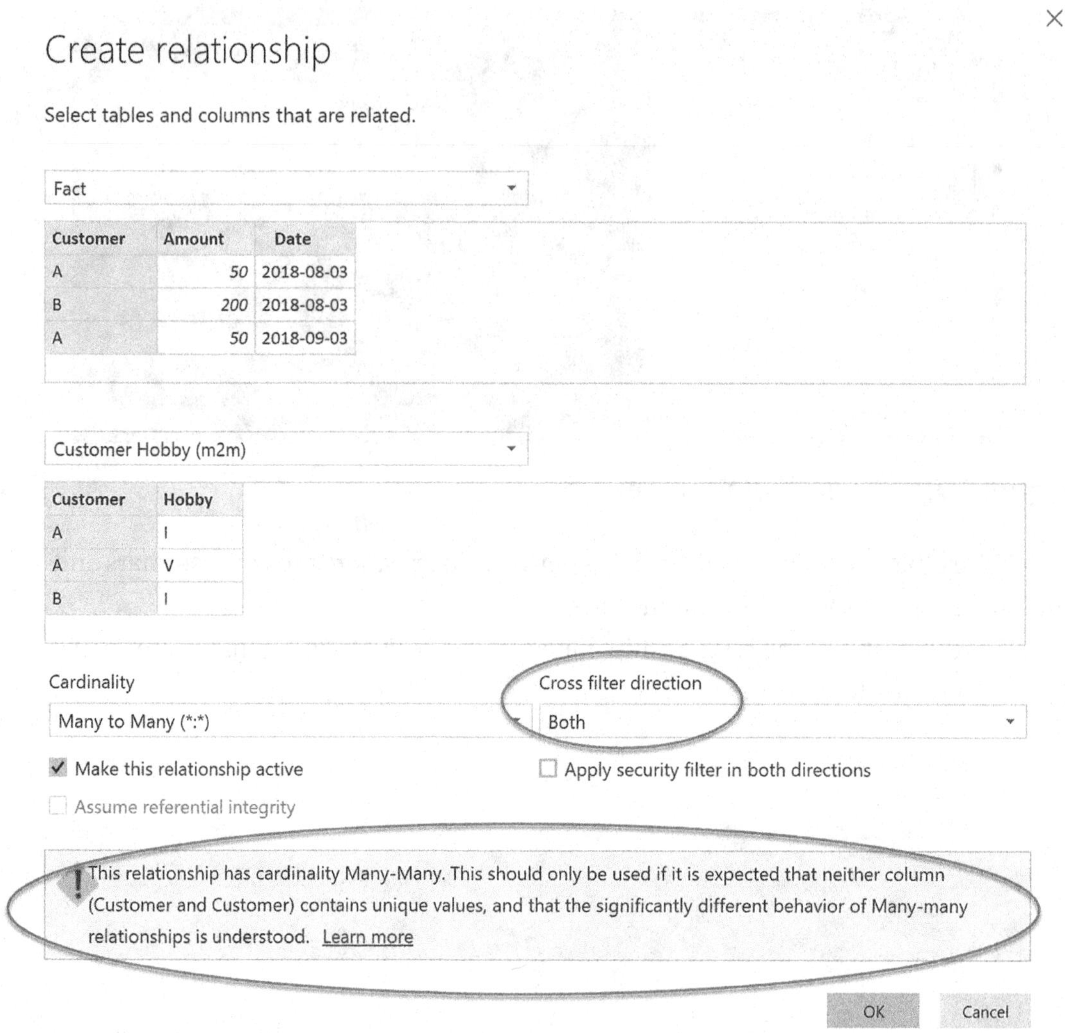

Create relationship   ×

Select tables and columns that are related.

Fact   ▼

Customer	Amount	Date
A	50	2018-08-03
B	200	2018-08-03
A	50	2018-09-03

Customer Hobby (m2m)   ▼

Customer	Hobby
A	I
A	V
B	I

Cardinality   Cross filter direction

Many to Many (*:*)   ▼   Both   ▼

☑ Make this relationship active   ☐ Apply security filter in both directions

☐ Assume referential integrity

! This relationship has cardinality Many-Many. This should only be used if it is expected that neither column (Customer and Customer) contains unique values, and that the significantly different behavior of Many-many relationships is understood. Learn more

                                                    OK   Cancel

Figure 2-20. *Relationships – many to many, cardinality*

As in the warning message stated, it's necessary to understand some significantly different behavior. This is

- RELATED no longer can be used

- ALL no longer removes the filter context

Next to different behavior, it's also necessary to change the value of the Cross filter direction. This is shown in Figure 2-21.

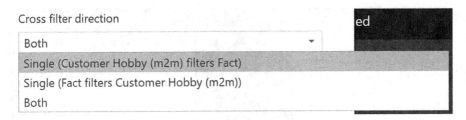

Figure 2-21. *Relationships – many to many, filter direction*

Figure 2-22 shows the result of the many-to-many relationship.

Customer Hobby (m2m) and Fact are directly related			
Customer	I	II	V	Total
A	100		100	100
B	200	200	200	200
Total	**300**	**200**	**300**	**300**

Figure 2-22. *Relationships – many to many, m2m result*

To avoid the different behavior, it's possible to create a model that just uses one-to-many relationships. This model is shown in Figure 2-23.

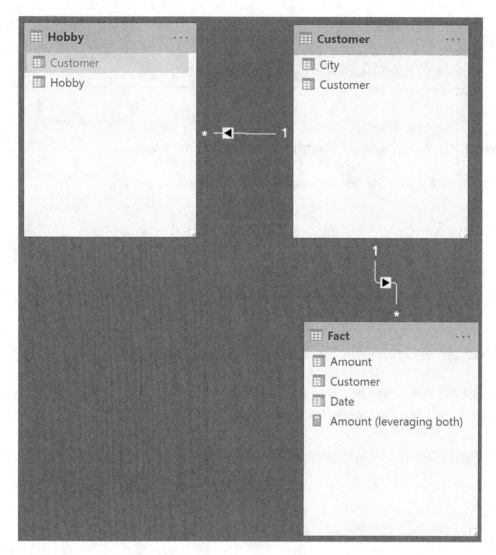

Figure 2-23. *Relationships – many to many, using one to many*

The result of this model is shown in Figure 2-24.

Using 1 to many relationships

Customer	I	II	III	IV	V	**Total**
A	100				100	**100**
B	200	200			200	**200**
Total	300	300	300	300	300	**300**

Figure 2-24. *Relationships – many to many, result of using one to many*

Both tables are showing the same results for each customer, but there is also a major difference between both tables. The table using the one-to-many relationships shows hobbies III and IV even if no such customer possesses corresponding rows in the fact table. This is simply due to the fact that the table Hobby is not directly related to the fact table.

This "issue" can be resolved by changing the Cross filter direction of the relationship between the tables Hobby and Customer from Single to Both. This is shown in Figure 2-25.

Cardinality	Cross filter direction
Many to one (*:1) ▾	Both ▾
☑ Make this relationship active	☐ Apply security filter in both directions

Figure 2-25. *Relationships – many to many, using one to many, Both*

Using relationships with a value of Both can become dangerous, because they can become part of a circular dependency if the data model becomes larger with a lot of tables.

For this reason, it's a better idea to use some DAX. This allows using just one-to-many relationships but still being able to avoid the NULL values. This can be achieved by using the DAX function CROSSFILTER. Listing 2-20 can be used to create this measure.

Listing 2-20. Relationships – using CROSSFILTER

```
Amount (leveraging cross filter direction both) =
CALCULATE(
    SUM('Fact'[Amount])
    , CROSSFILTER('Hobby'[Customer] , 'Customer'[Customer] , Both)
)
```

Whenever possible, it's recommended to use DAX instead of changing the Cross filter direction value to Both.

CHAPTER 3

DAX Lineage

Introduction

A handy concept to learn when working with DAX calculations is the concept of lineage. DAX lineage refers to the ability to trace the heritage of data in table expressions back to original physical tables. When new table expressions get built by various DAX functions, the columns in the new table *may* contain a reference back to the original physical table from which it was initially derived.

Lineage information can also get used by the DAX query engine to help decide what the most optimal approach is to produce the output in certain circumstances.

It's possible to write DAX calculations that include multigenerational statements in DAX that process, aggregate, and filter data from physical tables through a series of steps. These statements generate table expressions that get used for building other table expressions or used to filter or combine with other physical tables.

When the time comes for the DAX engine to process a calculation, many smart things happen under the covers, and knowing the lineage for columns enables the DAX query engine to take useful shortcuts to improve performance without compromising the accuracy of the outcome.

Note The DAX engine keeps track of column lineage for semantic reasons, not solely for performance optimization. The performance optimizer tracks other information which is related to column lineage, but not the same thing.

It is possible to write complicated DAX without thinking about lineage successfully, but knowing a little about what lineage is and how it works may at least explain why intellisense sometimes suggests columns as you type and other times it won't. Often this is because the intellisense can detect the status of lineage for the current position in the code.

© Philip Seamark, Thomas Martens 2019
P. Seamark and T. Martens, *Pro DAX with Power BI*, https://doi.org/10.1007/978-1-4842-4897-3_3

One way to think of lineage is as the family tree for every column in the model. At any one time, a column in a table expression knows if it is a descendant of royalty or native American chief. Most of the time it's not that exciting, and the column is usually the product of some unremarkable source system data. ☺

Definitions

A couple of essential definitions should get clarified before delving too far into lineage, and these are physical table and table expression.

Physical table

A physical table is any table in the database which you can directly query using the EVALUATE statement

```
EVALUATE 'Dates'
```

These tables are visible in the Report, Data, and Relationships views in Power BI Desktop. Alternatively, these tables can be hidden like the DAX automatically generated date tables.

Physical tables exist in memory and get considered as part of the overall footprint size of the data model.

Physical tables can be data imported into the model, or they can be created using the New Table button on the Power BI Desktop ribbon.

Columns in physical tables are assigned a unique column identifier by the DAX engine and do not have any lineage relationship to any other columns.

The DAX Storage Engine is the process that looks after retrieving data from physical tables.

Table expression

Table expressions, which also get referred to as virtual tables, are entities created temporarily and only exist for the short lifespan of a calculation. Table expressions are the result of any DAX function that returns a table as an output. These tables do not affect the overall memory footprint of the data model (at rest) but do consume memory at the point they are needed during the processing of a calculation.

When a table expression gets created by a DAX function from a physical table, the columns in the new table usually carry a reference to the unique column ID from the physical column from which it gets derived. This reference is stored internally and is not something you can easily see. This data is probably visible by querying the data model using DMV queries. Generally, this information is not useful for optimizing for performance.

The Logical or Physical query plans may provide a clue as to the column reference in their seemingly cryptic output, but again, it is not something likely to help when optimizing your calculations.

Example 1

Let's start with a simple example of a DAX function that produces a table expression and then walk through the components.

Consider the following calculated measure in Listing 3-1 which returns a single value that represents the total number of rows in the table expression generated by the SUMMARIZECOLUMNS function.

Listing 3-1. A simple example of a table expression

```
My Calculated Measure =
VAR MyTableExpression1 =
    SUMMARIZECOLUMNS(
            'Dimension Date'[Calendar Month Label] ,
            'Dimension Date'[Calendar Year Label] ,
            "Sum of Quantity" ,
            SUM('Fact Sale'[Quantity])
            )
RETURN
    MAXX(MyTableExpression1,[Sum of Quantity])
```

Note   Normally, you should not use SUMMARIZECOLUMNS in measure definitions. The example in Listing 3-1 is used to demonstrate table expressions.

In this example, the SUMMARIZECOLUMNS function returns a three-column table that gets assigned to the *MyTableExpression1* variable. The table generated by the SUMMARIZECOLUMNS function is not a physical table and only exists for the scope of the calculated measure. Once the calculated measure produces its final output, the table stored in the MyTableExpression1 variable is lost in time...like tears in the rain.

This temporary table assigned to the *MyTableExpression1* variable is a table expression. The table has three visible columns and probably a hidden column to keep track of each row identifier. Of the three visible columns, the first is a grouping of the 'Dimension Date'[Calendar Month Label] column. The syntax used in Listing 3-1 means the new column is created using the same name as the column in which it gets aggregated. Importantly, the new column also stores a reference back to the original physical column from where it gets sourced. In this case, the new column is only one step removed from the original column.

Renaming a column in a table expression does not break lineage references back to the original column. This metadata reference stored inside the table is the key to lineage.

The second column in the table expression called [Calendar Year Label] behaves in much the same way as the first. A reference back to the internal ID of the original column in the 'Dimension Date' table gets stored in the table expression as metadata.

The last visible column in the table expression is called "Sum of Quantity" and does not store an internal reference to the 'Fact Sale'[Quantity] column. This column is considered an original column regarding lineage and no longer has any direct relationship to the 'Fact Sale'[Quantity] column.

Finally, the calculated measure in Listing 3-1 passes the table expression stored in the *MyTableExpression1* variable to a MAXX function which iterates the number of rows to determine to a single value which represents the largest number across the Month/Year grouping. The result using the Wide World Importers dataset is 270,036.

The result means there are 41 distinct combinations of [Calendar Month Label] and [Calendar Year Label] in the dataset. This calculation measure does nothing with the sum total aggregation over the [Quantity] column.

The COUNTROWS function introduces a second layer of aggregation to the calculation when it reduces the number of rows in the table expression down to a single row. When looking at the profiler trace events for this calculation, it only produces a single Storage Engine (SE) scan. Figure 3-1 shows the Server Timings from this calculation in DAX Studio. This figure confirms only a single SE call gets made to the underlying database.

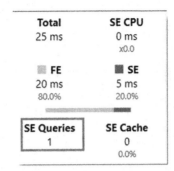

Figure 3-1. *Shows Server Timings for the query in Listing 3-1*

Despite the DAX logic suggesting two table-based operations should logically take place, only a single scan was needed to retrieve data to satisfy the output of the calculation. In this case, when the DAX engine comes up with a plan on how to satisfy the query, one factor taken into account is lineage-based relationships between every column used anywhere in the query, and where possible, the engine produces an efficient plan to make the most of the single storage engine call.

Example 2

Another way to look at how DAX lineage works is by using a table expression as a filter over another table.

Consider the calculation in Listing 3-2. This calculation intends to produce a list of any day from the 'Dimension Date' table that belongs to the best five months regarding quantity sold each month. The first step is to identify the best five months in the data, and then go back to the database and retrieve data for every date belonging to any of those five months.

Listing 3-2. A calculation using a table expression as a filter to highlight lineage

```
EVALUATE

VAR MyTableExpression1 =
    SUMMARIZECOLUMNS(
            'Dimension Date'[Calendar Month Label] ,
            "Sum of Quantity" ,
            SUM('Fact Sale'[Quantity])
            )
```

```
VAR myTableExpression2 =
    TOPN(
         5,
         MyTableExpression1,
         [Sum of Quantity]
         )

RETURN
    CALCULATETABLE(
         'Dimension Date',
         myTableExpression2
         )
```

The calculation in Listing 3-2 is a multistep process that first creates a table expression summarizing the 'Dimension Date' table down to calendar month. The SUMMARIZECOLUMNS function returns a table expression that has 41 rows and 2 columns, with every calendar month in the Wide World Importers dataset represented. The output of the SUMMARIZECOLUMNS function is stored in the *MyTableExpression1* variable and is not yet filtered to identify the top five months.

The *MyTableExpression2* variable stores the output of the TOPN function, which in this case uses the table expression created by the statement before its first parameter. The following parameters instruct the TOPN function to find the top five rows from the table expression when ordered by the [Sum of Quantity] column.

This new table expression now only has five rows but keeps the same two columns. Importantly, the [Calendar Month Label] column in the new table expression contains a lineage reference back to the original [Calendar Month Label] column in the 'Dimension Date' table. This reference becomes useful at the next step when it provides a filtering effect.

The final step in the query is to display all rows in the 'Dimension Date' table that belong to any of the top five months. The CALCULATETABLE function accepts table expressions as filters, and in this case, the first parameter of the CALCULATETABLE function says we should start with the entire 'Dimension Date' table that includes every column and all 1,461 rows.

The second parameter is where the magic of lineage happens. The *MyTableExpression2* variable contains a table expression with two columns. The CALCULATETABLE function recognizes the table passed as a filter has a column that shares a common lineage with the table passed as the first parameter. Once common columns get identified, the engine then builds a list of values for each common column (there can be more than one), and these values get then used as a filter over the base table.

Note   The primary reason the DAX engine keeps track of column lineage is to build filters using table expressions.

In the example shown in Listing 3-2, two calls are made by the storage engine to the database.

The first call by the storage engine aligns to the SUMMARIZECOLUMNS function that stores output in the *MyTableExpression1* variable, and the pseudo-T-SQL captured in a profiler trace gets shown in Listing 3-3. This query does not have a WHERE clause and grabs all data. The statement does not specify a GROUP BY clause, but the existence of a SUM function in the SELECT statement implies the resultset gets grouped by the [Calendar Month Label] column. Note the pseudo-T-SQL text in Listing 3-3 also includes reference numbers as part of table and column names.

Listing 3-3. The VertiPaq scan details for the first part of the query in Listing 3-2

```
SELECT
    [Dimension Date (12)].[Calendar Month Label (130)]
        AS [Dimension Date (12)$Calendar Month Label (130)],
    SUM([Fact Sale (1113)].[Quantity (1150)]) AS [$Measure0]
FROM [Fact Sale (1113)]
    LEFT OUTER JOIN [Dimension Date (12)]
        ON [Fact Sale (1113)].[Invoice Date Key (1130)]
            =[Dimension Date (12)].[Date (117)];
```

The second call made by the storage engine to the database gets shown in Listing 3-4. In the pseudo-T-SQL-like example, 14 individual column names have been removed to fit the text onto a single page. The important aspect is this: the query has a WHERE clause with a list of five values that get used as a filter over the [Calendar Month Label] column in the 'Dimension Date' table. These are dynamically created to reflect the top five months as identified by the previous step.

The original DAX code in Listing 3-2 never specifies which column filters should get applied. This filtering is taken care of by the DAX engine recognizing that a column used in the filter table in the filter argument of the CALCULATETABLE included a lineage reference column that matched a column in the first parameter. The DAX engine then derives the shared values between these columns to then build an IN statement that gets used by the second call by the storage engine to the database.

Listing 3-4. The VertiPaq scan details for the second part of the query in Listing 3-2

```
SELECT
    <... all fourteen column names removed for brevity ...>
FROM [Dimension Date (12)]
WHERE
    [Dimension Date (12)].[Calendar Month Label (130)] IN
    (
            'CY2016-Apr',
            'CY2016-Mar',
            'CY2015-Apr',
            'CY2016-May',
            'CY2015-Jul'
    );
```

The result is a 153-row table that includes all 14 columns from 'Dimension Date'. The 153 rows represent every day that belongs to any of the best five months based on quantity. They are not the best 153 rows from anywhere in the data table.

Renaming columns

The filtering in the CALCULATETABLE in Listing 3-2 works because of the lineage reference stored in the table expression. The same behavior gets produced even if column names get changed. The example in Listing 3-5 demonstrates lineage filtering is working despite column name changes along the way.

Listing 3-5. A modified version of the query in Listing 3-2 to demonstrate renaming columns on DAX lineage

```
EVALUATE

VAR MyTableExpression1 =
    SUMMARIZECOLUMNS(
        'Dimension Date'[Calendar Month Label] ,
        "Sum of Quantity" ,
        SUM('Fact Sale'[Quantity])
        )

VAR myTableExpression2 =
    SELECTCOLUMNS(
        TOPN(
            5,
            MyTableExpression1,
            [Sum of Quantity]
            ),
        "My New Columnname",
        [Calendar Month Label]
        )
RETURN
    CALCULATETABLE(
        'Dimension Date',
        myTableExpression2
        )
```

The TOPN function now has a SELECTCOLUMNS function wrapped around it, which is used to effectively rename the [Calendar Month Label] column to [My New Columnname]. The renamed column in the table expression generated by the SELECTCOLUMNS function still contains a reference back to the original column in 'Dimension Date'. The updated calculation in Listing 3-5 produces the same result as the calculation in Listing 3-2. Changing the name of the [Calendar Month Label] column does not break the filtering effect in the CALCULATETABLE function.

Figure 3-2 shows the multigenerational table expressions spawned during the calculation with arrows showing a link between a column in a table expression and a column in the original physical table.

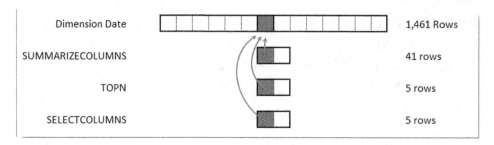

Figure 3-2. *Three table expressions including lineage arrows back to their original column*

Lineage and row context

Lineage ultimately determines how filtering works, as shown in the previous examples. If lineage is confirmed, then filtering takes place. If lineage cannot get traced, then the column does not use or get used for filtering.

DAX lineage also plays a crucial role in context transition.

Consider the following calculation in Listing 3-6 that returns one row for each unique value in the [Calendar Year] column. The example in Listing 3-6 happens to have four rows. Then in a second column, a value is calculated for each year to produce a value that represents the sum total of the [Quantity] column from 'Fact Sale'.

Listing 3-6. A simple example of a context transition to demonstrate the role of lineage

```
EVALUATE
    ADDCOLUMNS(
        VALUES('Dimension Date'[Calendar Year]) ,
        "Sum of Quantity" ,
        CALCULATE(
            SUM('Fact Sale'[Quantity])
            )
        )
```

The result of the query in Listing 3-6 gets shown in Figure 3-3. Each year has a different value in the [Sum of Quantity] column.

In the calculation in Listing 3-6, the VALUES function outputs a single column table expression with lineage back to the 'Dimension Date'[Calendar Year] column. The CALCULATE function clones the current filter context and brings through any row filtering in effect.

The first row of the table expression in the base table used in the ADDCOLUMNS table has a filter that [Calendar Year] should be 2013. The lineage details for this filtered column are passed through to the filter context created by the CALCULATE statement and applied to *any* column found that shares the same lineage.

As it happens, the extended 'Fact Sale' table already includes every column from the 'Dimension Date' table (see Chapter 5, "Filtering in DAX"), and because a column with a common lineage gets found, a filter gets placed over the column to only consider rows that equal 2013.

If the base table used for the first parameter of ADDCOLUMNS contained additional columns, the lineage details for these would also get passed to the CALCULATE function. Any common columns found based on lineage information in the extended 'Fact Sale' table get filtered accordingly.

Dimension Date[Calendar Year]	[Sum of Quantity]
2013	2401657
2014	2567401
2015	2740266
2016	1241304

Figure 3-3. *The result of the query in Listing 3-6*

Breaking lineage

Maintaining lineage through a series of table expressions can be handy, but sometimes you want to break lineage. Breaking lineage can be achieved using functions such as SELECTCOLUMNS and using the <expression> that doesn't change the value but is sophisticated enough for the function to decide not to add a lineage reference to the original column.

Consider the calculation in Listing 3-7 which extends the calculation in Listing 3-6. In this example, the VALUES function gets encompassed inside a SELECTCOLUMNS function. The SELECTCOLUMNS function receives the four-row, single-column table expressions from the VALUES function.

At this point, the column in this table expression contains a lineage reference back to the original column in 'Dimension Date'.

However, the SELECTCOLUMNS function creates a new table expression, again with a single column, but the DAX expression used to create the value for each row contains a + 0 in the statement. This expression is enough for the SELECTCOLUMNS function to produce a four-row, single-column table expression still, but now decides not to include a reference back to the original physical column. Thus, lineage now gets broken.

Listing 3-7. A modified statement demonstrates how to break lineage when needed

```
EVALUATE
    ADDCOLUMNS(
        SELECTCOLUMNS(
            VALUES('Dimension Date'[Calendar Year]) ,
            "Calendar Year" ,
            [Calendar Year] + 0
            ),
        "Sum of Quanity" ,
        CALCULATE(
            SUM('Fact Sale'[Quantity])
            )
        )
```

The CALCULATE statement still clones the existing filter context and passes row-level filtering information from the base table of the ADDCOLUMNS function, but as no common columns are found between the base table of ADDCOLUMNS and with any of the columns in the extended 'Fact Sale' table, the result gets shown in Figure 3-4.

[Calendar Year]	[Sum of Quanity]
2013	8950628
2014	8950628
2015	8950628
2016	8950628

Figure 3-4. *The result of the query in Listing 3-7*

Despite still naming the column "Calendar Year" in the SELECTCOLUMNS function, this does not have any effect. The lack of effective filters over the extended 'Fact Sale' table means each row in the second column produces the same value over and over, which happens to be the total for the entire column.

In this example, the lineage got broken because a zero gets added to it. For lineage to get included in the output, the syntax would need to follow the T[C] notation (which means 'tablename'[Column Name]), with no modifications. In this example, a zero gets used because the [Calendar Year] column is numeric and is enough to break the lineage – but not to change the value. If the core column uses a datatype of text, then appending an empty string has the same effect.

Faking lineage

A fantastic function added to the DAX language in 2018 is the TREATAS function. Among other things, this function specifically overwrites lineage details in table expressions. This lineage overwriting functionality is a bit like me having the ability to convince you that my great-grandfather was the English king, Henry VIII. Of course, he wasn't, but if you were a downstream DAX function, I could use the TREATAS function to fool you into thinking I genuinely originated from the famous king.

SYNTAX: TREATAS (<table> , <column> [,<column>]…)

OUTPUT: <table expression>

The syntax for TREATAS requires a table as the first parameter. This table can either be a physical table or a table expression. Then, as many <column> parameters must get supplied as there are columns in the table passed as the first parameter.

If the table passed as the first parameter of TREATAS has four columns, there need to be four parameters passed in addition to the table parameter. The order of the parameters gets matched to the order of the columns in the <table> passed, and they need to be of the same datatype.

Consider the calculation in Listing 3-8 that outputs a table designed to summarize the [Quantity] column down to the year. The DEFINE section of the calculation stores a table expression in a variable that is intended to get used as a filter for the CALCULATETABLE function. The DATATABLE function allows you to create a table expression using hardcoded values. In this case, the table expression stored in the *MyYearFilter* variable has a single column and only two rows and gets seen in Figure 3-5.

Importantly, this table expression has no lineage information linking it back to any physical table in the data model. Usually, this lack of any useful lineage in the table expression would make it useless as a filter table.

[Year]
2015
2016

Figure 3-5. *Shows the contents of the MyYearFilter variable in Listing 3-8*

Listing 3-8. A modified statement demonstrates how to break lineage when needed

```
DEFINE
    VAR MyYearFilter =
        DATATABLE(
            "Year" , INTEGER ,
            {
                    {2015},
                    {2016}
            }
        )
```

```
MEASURE 'Fact Sale'[Sum of Quantity] =
      SUM('Fact Sale'[Quantity])

EVALUATE
    CALCULATETABLE(
        SUMMARIZECOLUMNS(
            'Dimension Date'[Calendar Year],
            "Sum of Quantity",
            [Sum of Quantity]
            ),
        MyYearFilter
        )
```

In this example, the *MyYearFilter* table expression gets passed to the CALCUATETABLE as a filter parameter. The lineage details of every column in the table expression then get compared with every column in the extended version of 'Dimension Date'. With the code as it is, there are no common columns found, so the *MyYearFilter* table expression has no filtering effect over the SUMMARIZECOLUMNS column. The output for this code in Listing 3-8 gets shown in Figure 3-6.

Dimension Date[Calendar Year]	[Sum of Quantity]
2013	2401657
2014	2567401
2015	2740266
2016	1241304

Figure 3-6. *The results of the query in Listing 3-8*

The results in Figure 3-6 show even though the *MyYearFilter* variable gets passed to the CALCULATE function, it has no filtering effect, and the resultset still produces a value for [Calendar Year] 2013 and 2014. The final table was not filtered to just two years because the filter table expression shares no lineage with the table getting filtered.

The modification required to make the table expression assigned to the *MyYearFilter* variable work as a filter is to convince the CALCULATETABLE function that the columns in the *MyYearFilter* table expression genuinely originated from the 'Dimension Date' table. One way to achieve this is to store a table expression in *MyYearFilter* using DAX that uses the 'Dimension Date' physical table and therefore passes on lineage references. A version of this approach gets shown in Listing 3-9.

Listing 3-9. A modified table expression that uses unbroken lineage to filter the 'Dimension Date' table effectively

```
VAR MyYearFilter =
    FILTER(
        VALUES('Dimension Date'[Calendar Year]) ,
        [Calendar Year] IN {2015,2016}
        )
```

This can also be achieved using the TREATAS function shown in Listing 3-10.

Listing 3-10. A modified EVALUATE statement from Listing 3-8 that uses TREATAS to update the lineage details for the MyYearFilter table expression

```
EVALUATE
    CALCULATETABLE(
        SUMMARIZECOLUMNS(
            'Dimension Date'[Calendar Year],
            "Sum of Quantity",
            [Sum of Quantity]
            ),
        TREATAS(
            MyYearFilter ,
            'Dimension Date'[Calendar Year]
            )
        )
```

In the example in Listing 3-10, the TREATAS function gets passed to a *MyYearFilter* variable that contains a single-column table expression with only two rows. Because this table expression has one column, only one additional parameter needs to get passed to the TREATAS function. In this example, the 'Dimension Date'[Calendar Year] column gets specified.

The TREATAS function outputs a table expression with a single column that now includes a lineage reference to the 'Dimension Date'[Calendar Year] table, even though the table expression did not get generated from 'Dimension Date'.

The CALCUATETABLE now filters the SUMMARIZECOLUMNS using lineage from columns in the table filter parameter. A match now gets found, so the DAX expression in the first parameter applies a filter over the 'Dimension Date'[Calendar Year] column in the SUMMARIZECOLUMNS function using values from the column in the *MyYearFilter* table expression.

The new result gets shown in Figure 3-7. This result would be the same for both the FILTER and TREATAS methods shown in Listing 3-9 and Listing 3-10.

Dimension Date[Calendar Year]	[Sum of Quantity]
2015	2740266
2016	1241304

Figure 3-7. *The result of the updated query from either Listing 3-9 or Listing 3-10*

The technique of creating a small table expression of hardcoded values and using the TREATAS function to assign lineage details is used by Power BI Desktop when creating DAX for visuals on a canvas.

The example shown in Listing 3-8 used the DATATABLE function to generate a new table expression. This table could also have been generated using the following shorter syntax shown in Listing 3-11.

Listing 3-11. Shows an abbreviated method to create hardcoded table expressions

```
TREATAS(
    {2015,2016},
    'Dimension Date'[Calendar Year]
    )
```

In Listing 3-11, the first parameter of the TREATAS function uses a series of squiggly brackets and values to generate a table. The outer pair of {} brackets defines the overall table. This method only supports a single column that has the default column name of "Value" and cannot be used to generate multi-column tables.

Fixing broken lineage

If you recall the example back in Listing 3-7, the code was designed to highlight how it is possible to break lineage. In this case, a zero got added to an integer column so the DAX function would output a table expression without a lineage reference back to its source column.

The example in Listing 3-12 takes the code from Listing 3-7 and shows how the TREATAS function can get used to repair broken lineage.

Listing 3-12. Shows lineage restored using the TREATAS function based on code from Listing 3-7

```
EVALUATE
    ADDCOLUMNS(
        TREATAS(
            SELECTCOLUMNS(
                VALUES('Dimension Date'[Calendar Year]) ,
                "Calendar Year" ,
                [Calendar Year] + 0
                ),
            'Dimension Date'[Calendar Year]
            )    ,
        "Sum of Quantity" ,
        CALCULATE(
            SUM('Fact Sale'[Quantity])
            )
        )
```

In this example, the VALUES function outputs a single-column table expression. The single column in this table expression has a lineage reference back to the source column.

In the next step, the SELECTCOLUMNS function takes the table expression returned from the VALUES function and effectively breaks the lineage by adding a zero to the column.

Finally, the TREATAS function accepts the single-column table expression returned by SELECTCOLUMNS and assigns the column with a reference back to its source.

The CALCULATE function uses the repaired lineage when building the filter context for the SUM function. The results for the query in Listing 3-12 get seen in Figure 3-8.

Dimension Date[Calendar Year]	[Sum of Quantity]
2013	2401657
2014	2567401
2015	2740266
2016	1241304

Figure 3-8. *The results of the query from Listing 3-12*

Now ordinarily, you wouldn't write a query to deliberately break lineage only to restore it to what it previously was. This "wax on, wax off" example is designed to demonstrate the concept of how column lineage can get broken as well as reassigned in the spirit of the teaching of Mr. Miyagi. Mastering lineage moves you closer to writing the crane kick of DAX.

Summary

Understanding lineage and how it combines with filtering can take you a long way to solving difficult challenges using DAX. Knowing the difference between physical tables and table expressions can help explain why various functions in DAX behave the way they do. This chapter has shown not only how lineage works but how column lineage can be deliberately broken or reassigned as needed.

PART II

Core Concepts

CHAPTER 4

This Weird Context Thing

Explaining the context: Another approach

Much has been written, and much more has been said, about the evaluation context which has to be mastered to write the most complex DAX statements possible. And it will be very likely that this book will not be the last book that is written about DAX. This is due to the ongoing enhancements of this query language, and it's impossible to cover all the details about DAX or provide an exhaustive list of examples that will explain the application of DAX for each area of interest in just one book.

And to be clear, the first problem that requires some advanced DAX does not wait until all the things that could be learned are learned. It is more than likely that a current problem requires more complex DAX than the one shown in Listing 4-1.

Listing 4-1. The simple measure

```
the simple measure = CALCULATE(SUM('Fact Sale'[Quantity]))
```

Please, even if you are tempted to write a measure as shown in the preceding listing, do not do that. Using `CALCULATE` wrapped around an expression without using any additional parameters, meaning without using any filter expression, is considered a bad practice. Saying this, instead the preceding measure should look like as shown in Listing 4-1a.

Listing 4-1a. The simple measure

```
Measure with Filter Context = SUM('Fact Sale'[Quantity])
```

The simple truth is the questions we have to answer and the questions that are needing more complex DAX are not waiting until we feel well prepared. For this reason, we have to be brave and start writing DAX and solve real-world problems. But what if we don't have real-world problems?

© Philip Seamark, Thomas Martens 2019
P. Seamark and T. Martens, *Pro DAX with Power BI*, https://doi.org/10.1007/978-1-4842-4897-3_4

Joining the awesome people at community.powerbi.com and answering the questions asked is a great practice. Downloading the sample files that are provided and not being hesitant to ask if we do not fully understand the problem is an awesome learning experience. To become a DAX black belt, a DAX ninja, or whatever we want to become means to practice, practice, practice …

The previous chapters can be considered as a preliminary of a journey toward mastering advanced DAX. Consider the data model as an integral part of the analytical solution. Understanding DAX lineage is key to writing DAX statements and understanding why sometimes things do not work out as planned.

Whenever a DAX statement is executed, this execution happens within a context. This context determines the result of the DAX statement. For this reason, this context is called an evaluation context. This is because prior to the execution of the statement, the context is evaluated. Two different types of evaluation context have to be differentiated:

- Filter context

    This is the context that is evaluated whenever a DAX statement is used to calculate a measure. It's called filter context because before the expression is evaluated/executed, the implicit filters are applied, like selections from row and/or column headers, but also from slicers and the axis of visualizations.

- Row context

    This is the context that is implicitly present whenever a DAX statement is used to create a Calculated Column or if the expression is evaluated inside a table iterator function like SUMX(<table>, <expression>). The expression of the iterator function is executed in a row context.

Filter and row contexts: A gentle approach

In a way, it's easy to determine if there is a filter context or a row context. Let's assume we want to define a measure that simply aggregates the Quantity column from the Fact Sale table. Figure 4-1 shows that the DAX editor inside Power BI does not offer the selection of this column.

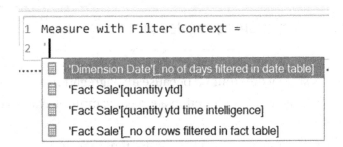

Figure 4-1. *Filter context – missing column*

This is simply because within a filter context, the reference to a column has to be wrapped into an aggregation function like SUM(<column reference>). Figure 4-2 shows the measure definition, but this time the DAX function SUM(…) is already used.

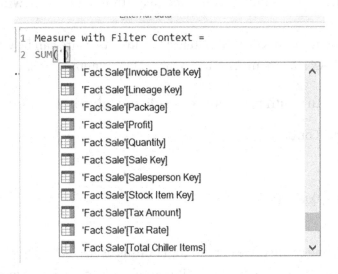

Figure 4-2. *Filter context – aggregation function*

Listing 4-2 shows the measure definition.

Listing 4-2. Filter context with the aggregation function

```
Measure with Filter Context =
SUM('Fact Sale'[Quantity])
```

Besides the obvious that now the column can be selected, there is something more that can be deduced from this fact.

Words of wisdom   A measure aggregates one or more rows.

This conclusion is simply determined because an aggregation function has to be used. This is because an aggregation function aggregates multiple values coming from different rows.

Of course, it's also possible to write a DAX statement that simply does not aggregate a single row. The preceding words of wisdom should be considered as a rule. The awesome Mr. Miyagi once said in the film *Karate Kid*

First learn stand, than learn fly. Nature rule Daniel-san, not mine.

Contemplating the rows being aggregated helps tremendously in writing DAX statements.

But unfortunately, or maybe fortunately, the same measure can be written using a slightly different DAX statement. This alternative DAX statement is shown in Listing 4-3.

Listing 4-3. Measure with row context inside the table iterator function

```
Measure with Row Context =
SUMX(
    'Fact Sale'
    ,'Fact Sale'[Quantity]
)
```

In Listing 4-3, one of the table iterator functions is used – SUMX (much more about iterator functions is written in Chapter 6, "Iterators"). What can be learned from Listing 4-3 is the following: whenever a row context is present, it's possible to reference a column without wrapping an aggregation function around the column. As Figure 4-3 shows, both measures return the same result.

City	Measure with Filter Context	Measure with Row Context
Abbottsburg	17,359	17,359
Absecon	12,415	12,415
Accomac	16,472	16,472
Aceitunas	12,693	12,693
Airport Drive	16,445	16,445
Akhiok	30,999	30,999
Alcester	12,802	12,802
Alden Bridge	14,645	14,645
Alstead	12,073	12,073
Amado	14,722	14,722
Amanda Park	12,221	12,221
Andrix	14,664	14,664
Annamoriah	15,326	15,326
Antares	15,363	15,363
Antonito	12,873	12,873
Arbor Vitae	14,334	14,334
Argusville	16,674	16,674
Total	**8,950,628**	**8,950,628**

Figure 4-3. *Filter context and row context side by side*

As soon as we are writing both these simple measures, we are instantaneously confronted with the evaluation context. We have to use an aggregation function whenever we are using a column reference like `'Fact Sale'[Quantity]` within an existing filter context, but we can omit this aggregation function by just referencing the column inside the table iterator. This is due to the row context that is created by the table iterator functions.

As can be seen, there is more than one way to define a measure. This is of course a good thing. But this possibility also comes with a burden. As there are more solutions for the same problem, the writing of DAX statements can become complex. This is especially the case when the datasets and the number of rows being aggregated are growing.

Filter and row contexts: Maybe some weird observations

Maybe you are totally aware of the content from the last chapter. However, consider the preceding text as a recap. To add some complexity, let's have a look at the measure shown in Listing 4-4.

Listing 4-4. Measure with row context and using SUM

```
Measure with Row Context and Sum =
SUMX(
    'Fact Sale'
    ,SUM('Fact Sale'[Quantity])
)
```

In comparison to Listing 4-3, the statement from Listing 4-4 just wraps the aggregate function SUM around the column reference. The result of the measure `Measure with Row Context and Sum` is shown in Figure 4-4.

City	Measure with Filter Context	Measure with Row Context	Measure with Row Context and Sum
Abbottsburg	17,359	17,359	7,221,344
Absecon	12,415	12,415	3,960,385
Accomac	16,472	16,472	6,061,696
Aceitunas	12,693	12,693	4,366,392
Airport Drive	16,445	16,445	6,643,780
Akhiok	30,999	30,999	22,691,268
Alcester	12,802	12,802	4,544,710
Alden Bridge	14,645	14,645	5,198,975
Alstead	12,073	12,073	3,802,995
Amado	14,722	14,722	5,579,638
Amanda Park	12,221	12,221	4,265,129
Andrix	14,664	14,664	5,220,384
Annamoriah	15,326	15,326	5,593,990
Antares	15,363	15,363	5,668,947
Antonito	12,873	12,873	4,595,661
Arbor Vitae	14,334	14,334	4,930,896
Argusville	16,674	16,674	6,369,468
Total	**8,950,628**	**8,950,628**	**2,043,115,100,420**

Figure 4-4. *Measure with row context and SUM*

Before the result of this measure will be explained in detail, a last measure will be introduced. The definition of this measure is shown in Listing 4-5.

Listing 4-5. Measure with row context using SUM and CALCULATE

```
Measure with Row Context and Sum and Calculate =
SUMX(
    'Fact Sale'
    ,CALCULATE(SUM('Fact Sale'[Quantity]))
)
```

The result of this measure is shown in Figure 4-5.

City	Measure with Filter Context	Measure with Row Context	Measure with Row Context and Sum	Measure with Row Context and Sum and Calculate
Abbottsburg	17,359	17,359	7,221,344	17,359
Absecon	12,415	12,415	3,960,385	12,415
Accomac	16,472	16,472	6,061,696	16,472
Aceitunas	12,693	12,693	4,366,392	12,693
Airport Drive	16,445	16,445	6,643,780	16,445
Akhiok	30,999	30,999	22,691,268	30,999
Alcester	12,802	12,802	4,544,710	12,802
Alden Bridge	14,645	14,645	5,198,975	14,645
Alstead	12,073	12,073	3,802,995	12,073
Amado	14,722	14,722	5,579,638	14,722
Amanda Park	12,221	12,221	4,265,129	12,221
Andrix	14,664	14,664	5,220,384	14,664
Annamoriah	15,326	15,326	5,593,990	15,326
Antares	15,363	15,363	5,668,947	15,363
Antonito	12,873	12,873	4,595,661	12,873
Arbor Vitae	14,334	14,334	4,930,896	14,334
Argusville	16,674	16,674	6,369,468	16,674
Total	**8,950,628**	**8,950,628**	**2,043,115,100,420**	**8,950,628**

Figure 4-5. *Measure with row context and SUM and CALCULATE*

Figure 4-5 shows that the measures

- `Measure with Row Context`

- `Measure with Row Context and Sum and Calculate`

using the iterator function SUMX return the same value as the more simple measure Measure with Filter Context. An explanation why this is the case for the measure Measure with Row Context already has been given. The reason is this: Inside a row context, the value from the referenced column for each row from the table that is iterated will be passed to the iterator function SUMX.

The reason why the measure that wraps the CALCULATE around the SUM creates the same result is the following: CALCULATE transforms the row context created by SUMX into a filter context. This means that each column value of the entire row that currently is iterated is transformed into a filter. The table Fact Sale then is filtered by this newly created filter. This transformation is called **context transition**.

Note   CALCULATE transforms an existing row context into a filter context.

As there is a unique identifier in the table Fact Sale, namely, the column Sale Key, it's safe to say that the current row from the iterator that is transformed into a filter is filtering itself.

Now to the measure Measure with Row Context and Sum, maybe it's not that obvious what's going on, as the values might look odd, and it's likely that the result is not what is expected. For this reason, Listing 4-6 shows the definition of this measure once again.

Listing 4-6. Measure with row context and using SUM

```
Measure with Row Context and Sum =
SUMX(
    'Fact Sale'
    ,SUM('Fact Sale'[Quantity])
)
```

The iterator function SUMX iterates over the table Fact Sale, but before the iteration starts, all existing filters are applied. Remember a measure is evaluated in a filter context. The filter in this simple example is the value from the row header coming from the column City of the table Dimension City. This means that the number of rows present in the table Fact Sale is filtered down to 416 rows for the city of Abbottsburg before the measure gets evaluated. This can be validated if the measure _no of rows filtered in fact table from Listing 4-7 will be added to the Matrix visual, this Matrix visual is shown in Figure 4-6.

Listing 4-7. Number of filtered rows of the table Fact Sale

```
_no of rows filtered in fact table =
COUNTROWS('Fact Sale')
```

City	Measure with Filter Context	Measure with Row Context	Measure with Row Context and Sum	Measure with Row Context and Sum and Calculate	_no of rows filtered in fact table
Abbottsburg	17,359	17,359	7,221,344	17,359	416
Absecon	12,415	12,415	3,960,385	12,415	319
Accomac	16,472	16,472	6,061,696	16,472	368
Aceitunas	12,693	12,693	4,366,392	12,693	344
Airport Drive	16,445	16,445	6,643,780	16,445	404
Akhiok	30,999	30,999	22,691,268	30,999	732
Alcester	12,802	12,802	4,544,710	12,802	355
Alden Bridge	14,645	14,645	5,198,975	14,645	355
Alstead	12,073	12,073	3,802,995	12,073	315
Amado	14,722	14,722	5,579,638	14,722	379
Amanda Park	12,221	12,221	4,265,129	12,221	349
Andrix	14,664	14,664	5,220,384	14,664	356
Annamoriah	15,326	15,326	5,593,990	15,326	365
Antares	15,363	15,363	5,668,947	15,363	369
Antonito	12,873	12,873	4,595,661	12,873	357
Arbor Vitae	14,334	14,334	4,930,896	14,334	344
Argusville	16,674	16,674	6,369,468	16,674	382
Total	8,950,628	8,950,628	2,043,115,100,420	8,950,628	228265

Figure 4-6. *The final Matrix visual*

Remembering that the iterator function SUMX creates a row context, the explanation for the odd result is this:

- SUMX iterates over the filtered table Fact Sale, already filtered by the implicit filters.

- SUM aggregates all the rows available.

The division of 7221344 by 416 results in 17359. This means the expression is executed 416 times. Each time, the aggregation function SUM is executed, aggregating the filtered rows. The result for each aggregation is then finally passed to SUMX.

If you have access to an SQL Server where the database Wide World Importers is installed, you can create the same result by using the SQL statement from Listing 4-8.

Listing 4-8. Filter context – iterator sumx

```
USE [WideWorldImportersDW]
GO
WITH
acte AS(
SELECT
    c.City
,   f.Quantity
```

```
,    SUM(f.quantity) OVER(PARTITION BY  c.[city])AS sumx_usingsum
,    SUM(f.quantity) OVER(PARTITION BY  f.[sale key])    AS sumx_
usingcalculateandsum
FROM
    Fact.Sale AS f
        INNER JOIN Dimension.City AS c ON
            f.[City Key] = c.[City Key]
)
SELECT
    acte.city
,    SUM(acte.quantity) AS FilterContext
,    SUM(acte.quantity) AS SUMX_JustColumnReference
,    SUM(acte.sumx_usingsum) AS SUMX_UsingSum
,    SUM(acte.sumx_usingcalculateandsum) AS
SUMX_UsingSumAndCalculate , COUNT(*) AS noofrows_FactSale
FROM
    acte
GROUP BY
    acte.city
ORDER BY acte.City ASC
```

Figure 4-7 shows a fraction of the result.

	city	FilterContext	SUMX_JustColumnReference	SUMX_UsingSum	SUMX_UsingSumAndCalculate	noofrows_FactSale
1	Abbottsburg	17359	17359	7221344	17359	416
2	Absecon	12415	12415	3960385	12415	319
3	Accomac	16472	16472	6061696	16472	368
4	Aceitunas	12693	12693	4366392	12693	344
5	Airport Drive	16445	16445	6643780	16445	404
6	Akhiok	30999	30999	22691268	30999	732
7	Alcester	12802	12802	4544710	12802	355
8	Alden Bridge	14645	14645	5198975	14645	355
9	Alstead	12073	12073	3802995	12073	315
10	Amado	14722	14722	5579638	14722	379

Figure 4-7. *Filter and row contexts – SQL statement*

If you are currently more fluent in speaking SQL than DAX, you can compare the DAX aggregation functions like SUM to the corresponding SQL functions of course in combination with the GROUP BY clause. The table iterator functions can be compared to the windowing functions in SQL following this pattern:

```
AGGREGATION(column) OVER(PARTITION BY ..., ... ORDER BY ..., ...)
```

A hint – just a hint

As we have seen in Listing 4-5, the use of the function CALCULATE fixed an issue, just by wrapping it around the aggregation function. As CALCULATE is a powerful function, it's also necessary to use this function consciously. This means whenever a row context has to be transformed, using CALCULATE is mandatory. If a filter context has to be changed to create the needed results, using CALCULATE is also necessary.

These are the only reasons for using CALCULATE:

- Transforming a row context into a filter context

- Changing an existing filter context

Don't get lost

Please consider the following: The VertiPaq engine is fast, because it's an in-memory columnar database. So whenever we write a DAX statement, we also have to answer the following question: Is the DAX statement as efficient as possible? The optimization of a DAX statement can become even more time consuming than the first version that just returns the correct results. Striving for the most efficient DAX statement can also mean an additional table has to be created (invisible for the common user of the data model) and one or more Calculated Columns have to be defined.

For this reason, a DAX statement can be considered as good enough if it returns the results without hindering the decision makers, the users of the DAX statement that we are creating, in their daily work. This approach can liberate some energy that can be spent on finding unique solutions that will provide tremendous business value. Because we do not constantly have to ask if we really know each detail, most of us do not need to create DAX statements that wade through billions or trillions of rows.

But we have to prepare for future data growth. For this reason, Chapter 14, "Scale Your Models," discusses how we can scale a model, of course using DAX. At the beginning of this chapter, it was said that *"And to be clear, the first problem that requires some advanced DAX does not wait until all the things that could be learned are learned."*

The following can be considered as a mantra that may help us refocus whenever we don't get the result that we are expecting.

It's just a single value, most of the time a number

Whenever a measure or Calculated Column has to be written, it has to be composed. We should never forget that most of the time the expected result of a DAX statement is a single value. Of course, we use DAX to create tables to filter other tables to filter other tables to filter other tables and so on ... But whenever this filtering is done, a single value is returned. Sometimes things happen, and it's just the last aggregation or just a SELECTCOLUMN to create a result that can be implicitly coerced to a scalar value.

The scalar value is an aggregation

Whenever we are creating a measure, we have to be aware that rows are aggregated. There is just one exception, we are creating a Calculated Column and do not use any aggregation function. This means it's not the expression that is responsible for unexpected results, but the number of rows, or we are just aggregating the wrong rows.

The aggregation is fed by filtered rows

Through the following chapters of this book starting with Chapter 5, "Filtering in DAX," you will learn how to filter the rows that will be aggregated. If it's not clear what this means, just consider these two simple measures:

- Quantity previous month

  If we are going to create a measure that returns the value of the previous month, we have to "shift" the filter of the current month (meaning the month that is currently filtering the table Fact Sale) to the previous month.

- City contribution to all cities

    Now we have to divide the value. Let's say the quantity sold in a selected city (either by a slicer or a row/column header) by the total value of all cities. To determine the value of all cities, the number of rows has to be expanded, meaning we have to remove the existing filter.

Each filter is a table

Even if we write a measure as in Listing 4-9

Listing 4-9. Even a simple filter is a table

```
Even a simple filter is a table =
CALCULATE(
    SUM('Fact Sale'[Quantity])
    ,'Dimension City'[City] = "Abbottsburg"
)
```

the Boolean filter expression ,'Dimension City'[City] = "Abbottsburg" will be internally "translated" into the statement as in Listing 4-10.

Listing 4-10. The well-formed scalar as expression as FILTER

```
The well-formed scalar as expression as FILTER =
CALCULATE(
    SUM('Fact Sale'[Quantity])
    ,FILTER(
        ALL('Dimension City'[City])
        ,'Dimension City'[City] = "Abbottsburg"
    )
)
```

Knowing this, start thinking in tables if you want to modify an existing filter context. This becomes much more evident after you read about extended columns in the next chapter.

If you got stuck, this trick may help: Create a single Excel sheet with all of the columns of your data model, at least the ones that you are using as slicers, row/column headers, or x-axis in charts. Use the table that is on the many side of the relationships and add the columns from the tables that are on the one side. Enable the filter function and select the values that are currently selected in your Power BI report. The filtered rows in the Excel sheet are representing the filtered fact table with all the extended columns from the related tables on the many side. Now, it is simple to discover how to expand or further reduce the rows in your DAX statement.

CHAPTER 5

Filtering in DAX

Introduction

Power BI does a fantastic job of allowing a report author to create a series of visuals that provide ways for an end user to explore a data model using a drag-and-drop approach. Sooner or later, there is a requirement that involves the same author adding some form of filtering to a DAX calculation.

This chapter explores and attempts to explain the concepts and hopefully clarifies a few nuances of the mysterious world of calculation-based DAX filtering along the way.

At first glance, filtering rules *appear* simple. The logic and code used is a series of expressions that resolve to a Boolean TRUE or FALSE to determine if a row is considered relevant for a given calculation.

The fun begins when you need to consider the order in which filters get applied. Filter rules get applied in layers and can overwrite and sometimes contradict each other in ways that can quickly become confusing when working on advanced calculations.

Once all layers get resolved, the rows deemed to have survived all filtering logic then get used by the core calculation to produce a result.

The basics

Adding filter criteria to a DAX calculation can take two forms. The first form uses column-based rules such as [State Province] = 'Texas' or [Calendar Year] = 2014. This approach specifies a value that is used to compare with values in a column. When statements like these get used, they are evaluated down to a Boolean True or False to determine if a row should be retained or discarded. A Boolean True means the row gets retained, while a Boolean False means the row should be ignored and not considered.

These types of statements can combine with other Boolean tests and follow standard Boolean logic when used with operators such as AND (&&) and OR (||).

© Philip Seamark, Thomas Martens 2019
P. Seamark and T. Martens, *Pro DAX with Power BI*, https://doi.org/10.1007/978-1-4842-4897-3_5

The second form of filtering uses DAX tables as filter parameters in a function that accepts tables as a filter. Passing a table as a filter argument has the effect of applying filtering across multiple columns, each with multiple values to determine the rows from a table that get considered.

Boolean filtering

The FILTER, CALCULATE, and CALCULATETABLE functions all accept filtering instructions in the format [Column] = <some value>. A basic example of Boolean filtering gets shown in Listing 5-1 and Listing 5-2. The CALCULATE and CALCULATETABLE functions convert what appears to be a Boolean statement into a format that returns a table expression to get used as a filter.

Listing 5-1. A basic example of Boolean filter expression in a FILTER function

```
COUNTROWS(
    FILTER(
        'Fact Sale',
        'Fact Sale'[Quantity] = 3
        )
    )
```

Listing 5-2. A basic example of Boolean filter expression in a CALCULATE function

```
CALCULATE(
    COUNTROWS('Fact Sale'),
    'Fact Sale'[Quantity] = 3
    )
```

The Boolean condition in Listings 5-1 and 5-2 applies to the distinct values of the [Quantity] column to determine if the row gets considered as part of the final total. If no other filters are applied, they both return the same value of 12,572, from an overall row count for the 'Fact Sale' table of 228,265.

Additional criteria can be added to the logic so long as the DAX expression can be evaluated down to a single TRUE or FALSE value. Brackets can be used to group Boolean logic gates to provide a wider variety of possibilities in more complex scenarios.

The effect of using brackets and Boolean operators is consistent with other languages and should feel familiar especially if you have worked with T-SQL-style predicates.

Note   While the calculations in Listings 5-1 and 5-2 appear the same, they yield different results when used in conjunction with external filters such as slicer selections and axis values in visuals.

When using the FILTER function, the first parameter required is a reference to the table to filter. This parameter can be a physical table or a table expression. Boolean expressions in the second parameter can use any column from the table specified in the first parameter. It's also possible to use columns in related tables with the help of the RELATED and RELATEDTABLE functions as shown in Listing 5-3.

Listing 5-3. A FILTER function using the RELATED function to apply filter across a relationship

```
COUNTROWS(
    FILTER(
        'Fact Sale',
        RELATED('Dimension City'[State Province]) = "Texas"
        )
    )
```

In Listing 5-3, the FILTER function uses the RELATED function to filter to a column in a related parent table. If no other external filters are applied, the result is 15,800 and represents the number of records in the 'Fact Sale' table sold in the [State Province] of "Texas".

The CALCULATE function applies Boolean filters in a different way that is explained in more detail later in this chapter that doesn't require the use of the RELATED and RELATEDTABLE functions to reference columns in related tables.

Tables as filters

The second way to provide filter criteria in DAX is in the form of a table. This approach can be used in the <filter> argument for both the CALCULATE and CALCULATETABLE functions and not only offers the opportunity to filter by a list of values but also affects multiple columns in a single step.

Note   Under the cover, the only filter supported by CALCULATE is table expression based. The Boolean syntax is internally converted to a table expression using the FILTER function.

If you are familiar with the concept of set-based programming in languages like T-SQL, you can consider this as a form of set-based *filtering*. Many filter instructions can get applied in a single step. Table filtering is quite powerful and involves a certain degree of mental gymnastics while learning the technique.

Single-column table filter

Consider the following example in Listing 5-4 that uses a single-column table expression to filter a calculation. The single-column table to be used as a table filter contains every [State Province] whose name begins with the letter N.

Listing 5-4. A CALCULATE function using a single-column table with multiple rows as a filter

```
Test Measure =
VAR MyFilterTable =
    FILTER(
        VALUES('Dimension City'[State Province]),
        LEFT('Dimension City'[State Province],1) = "N"
        )
RETURN
    CALCULATE(
        SUM('Fact Sale'[Quantity]),
        MyFilterTable
        )
```

The variable called *MyFilterTable* gets assigned a single-column table expression with nine rows, and the complete dataset gets shown in Figure 5-1.

Figure 5-1. *Shows contents of the MyFilterTable variable from Listing 5-4*

When a table gets passed to the CALCULATE (or CALCULATETABLE) function, it has the effect of filtering values that don't appear in this list in matching columns.

Several things to highlight in the calculation in Listing 5-4 are as follows.

Extended virtual columns

The core calculation in Listing 5-4 uses a SUM function over a column in the 'Fact Sale' table. In this calculation, the column in the table filter does not directly match with any native column in the 'Fact Sale' table. The filter that gets applied represents a column in a related table but does not use the RELATED function.

Filtering works here because a relationship exists between the 'Fact Sale' and 'Dimension City' tables using the [City Key] column from both tables.

When a standard one-to-many relationship gets defined in a model, the table that sits on the many side is theoretically extended to include all upstream tables on the one side, as virtual columns for calculations and filtering. In this case, the filter applied to the 'Dimension City'[State Province] behaves as if the data model gets flattened, and the 'Fact Sale' table physically includes this column.

Lineage

So how does the CALCULATE function know which column should be the subject of the filter in the table expression? The answer is not by matching column names between the table used as a filter and the table to be filtered.

Note   Columns in table expressions only filter physical columns that share a common lineage – Chapter 3, "DAX Lineage," has more details on lineage.

In the example in Listing 5-4, the *MyFilterTable* variable uses a FILTER function to reduce the number of [State Province] values down to only those that begin with the letter "N". The VALUES function creates a distinct list of every value in the 'Dimension City'[State Province] column.

The internal column ID reference for the 'Dimension City'[State Province] column gets stored in the table expression assigned to the *MyFilteredTable* variable. Then, when the table variable gets applied as a filter in the CALCULATE function, the internal column ID reference from the table expression is used to identify columns in the extended version of the 'Fact Sale' table which includes every column from the 'Dimension City' table.

Internal column ID references are unique to the database, and even if a column name gets changed by functions such as SELECTCOLUMNS or aggregated using functions like SUMMARIZE, if the reference ID survives through to the point the table expression gets applied by the CALCULATE function, the filter works as expected.

The following calculation in Listing 5-5 expands on the code in Listing 5-4 and highlights the effect of lineage on filtering.

Listing 5-5. A CALCULATE function using a single-column table with multiple rows as a filter

```
Test Measure 2 =
VAR MyFilterTable =
    SELECTCOLUMNS(
        FILTER(
            VALUES('Dimension City'[State Province]),
            LEFT('Dimension City'[State Province],1) = "N"
            ),
```

```
            "Something else",[State Province],
            "State Province","dummy value"
            )
RETURN
    CALCULATE(
        SUM('Fact Sale'[Quantity]),
        MyFilterTable
        )
```

Something else	State Province
N/A	dummy value
New York	dummy value
New Mexico	dummy value
North Dakota	dummy value
Nevada	dummy value
Nebraska	dummy value
North Carolina	dummy value
New Jersey	dummy value
New Hampshire	dummy value

Figure 5-2. Shows contents of the MyFilterTable variable in Listing 5-5

Note the original [State Province] column Shown in Figure 5-2 gets renamed and a new dummy column gets added with a meaningless hardcoded value that doesn't exist in the database. When this table expression gets applied as a filter to the 'Fact Sale' table by the CALCULATE function, the result is the same as for the calculation in Listing 5-4.

The internal column ID reference is used to match the list of values that get applied as filters to the table used in the <expression> of the CALCULATE table.

So even though the column is now called "Something else," the values in this column are still used to filter the physical table.

Multi-column table filter

A table expression used as a filter does not need only to have a single column. The example shown in Listing 5-5 uses a multi-column table as a filter argument. When a CALCULATE function is processing filter tables, it checks every internal column ID in the filter table and looks to see if there is a match in the table getting filtered. Every column that matches gets filtered, and unmatched columns from both tables get ignored.

In Listing 5-6, a table expression that includes two columns gets assigned to the *MyFilterTable* variable. A filter gets applied to the columns, so it only shows cities that start with the letter N located in states that also start with the letter N. This filter criterion is for demonstration purposes and unlikely ever to be a real-world example. The table expression stored in the *MyFilterTable* variable has 234 rows with a sample of the data shown in Figure 5-3.

In this example, the ALL function is used instead of the VALUES function. The ALL function returns a distinct list of values from the entire table that ignores any external filtering that may exist.

Listing 5-6. A CALCULATE function using a multi-column table with multiple rows as a filter

```
Measure 5-6 =
VAR MyFilterTable =
    FILTER(
        ALL(
            'Dimension City'[State Province],
            'Dimension City'[City]
            ),
        LEFT('Dimension City'[State Province],1) = "N" &&
        LEFT('Dimension City'[City],1) = "N"
        )
RETURN
    CALCULATE(
        SUM('Fact Sale'[Quantity]),
        MyFilterTable
        )
```

State Province ▾	City ▾
New York	Nanticoke
New York	Napoli
New York	Nashville
New York	Nassau Shores
New York	Nelson
New York	Neversink
New York	New Albion
New York	New Baltimore

Figure 5-3. *Shows a sample of contents of the MyFilterTable variable from Listing 5-6*

The CALCULATE function in Listing 5-6 is the same as the example in Listing 5-5. No information is needed to instruct the function on which columns the filter should be applied. The filters are applied based on the internal column ID.

The result is the CALCULATE function now applies filter logic using two columns from the extended 'Fact Sale' table. Both columns in the table expression passed as a table filter find matching columns in the extended 'Fact Sale' table. The SUM function returns a value of 85,999, but what does this mean?

Consider the example shown in Figure 5-4. The table on the left at step 1 is a table to be filtered by a multi-column table at step 2. Assume the lineage of the two columns in each table line up.

The result after filtering is complete is shown by the dataset at step 3. The example in Figure 5-4 shows filtering is not accumulative (Boolean AND); instead, it is based on an OR predicate. The two columns involved in the filter are tested independently of each other, and the outcome reflects the following logic:

Column1 IN ("A", "B") OR Column2 IN (1, 4)

If you were surprised to see the third row at step 3, you were probably expecting the test to get based on an AND test.

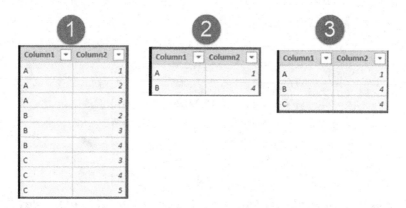

Figure 5-4. *Shows the result when the table on the left (1) gets filtered by the middle table (3). The result is the table on the right (3)*

Multiple filters

It's possible to provide multiple tables using the same technique. The CALCULATE and CALCULATETABLE functions can accept multiple filter parameters, meaning other table expressions can be used to extend the filtering logic to include columns from other tables. Listing 5-7 shows the syntax definition for the CALCULATE function.

Listing 5-7. Syntax signature for the CALCULATE function

```
CALCULATE(<expression>,<filter1>,<filter2>...)
```

The calculated measure in Listing 5-8 is similar to the calculated measure in Listing 5-6; however, a new table expression is added and assigned to the *MyFilterTableYear* variable. The new table variable gets added as the third parameter to the CALCULATE function in the RETURN statement. This additional filter has the effect of filtering the 'Fact Sale' table down to activity associated with the calendar year of 2014.

Listing 5-8. A CALCULATE function using multiple table expressions as a filter

```
Measure 5-8 =
VAR MyFilterTable =
    FILTER(
        ALL(
            'Dimension City'[State Province],
            'Dimension City'[City]
            ),
```

```
        LEFT('Dimension City'[State Province],1) = "N" &&
        LEFT('Dimension City'[City],1) = "N"
        )
VAR MyFilterTableYear =
    FILTER (
            'Dimension Date' ,
            'Dimension Date'[Calendar Year] = 2014
        )
RETURN
    CALCULATE(
        SUM('Fact Sale'[Quantity]),
        MyFilterTable,
        MyFilterTableYear
        )
```

A potential performance issue to consider in this example is the *MyFilterTableYear* variable stores every column from the 'Dimension Date' table and not just the [Calendar Year] column. There are 14 columns in the 'Dimension Date' table, so when the table expression gets applied in the CALCULATE function, filtering from all 14 columns gets processed which can result in a large amount of unnecessary processing.

A better approach would be to only store the bare minimum number of columns required to be used for filtering in the table expression. The DAX pattern used for the *MyFilterTable* variable is a more efficient example in that it only includes just the columns needed for filtering

Extended (virtual columns)

A key concept to understand for both filtering and evaluation context is the idea of extended virtual columns. Every time a one-to-many relationship gets defined in the data model, the table that sits on the many side is logically extended to include every column from the table on the one side for filtering and calculations. The extended table includes every column from any upstream table based on a many-to-one direction and can pass many generations.

Consider the set of tables shown in Figure 5-5. There are four tables:

1. Fact Sale – 21 columns

2. Dimension Date – 14 columns

3. Dimension Stock Item – 20 columns

4. Dimension City – 14 columns

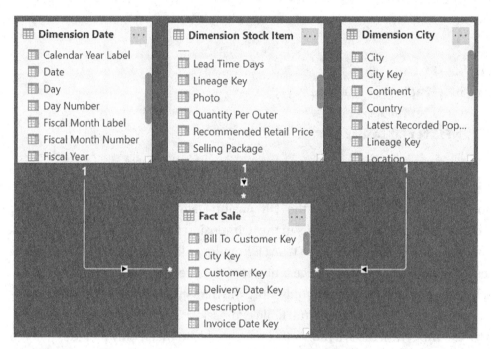

Figure 5-5. *Four tables from the WWI dataset in the Relationships view*

Any calculation based on the 'Fact Sale' table does not deal with a table that has only 21 rows; rather, the table is extended to include every column from related parent tables as well. If the data model included just the four tables as shown in Figure 5-5, the extended 'Fact Sale' table would have 69 columns, 21 native columns, and 48 extended columns.

The connection between the 'Fact Sale' table and the extended tables is considered a LEFT OUTER JOIN, which means that all rows from 'Fact Sale' but only rows from the extended tables that have a matching value in the column used to define the relationship are returned.

Figure 5-6 shows a widened 'Fact Sale' table that includes the additional columns from the extended tables.

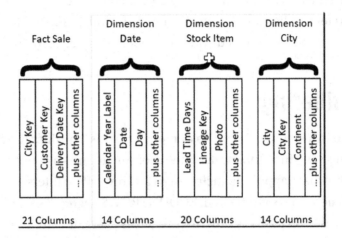

Figure 5-6. *Shows the Fact Sale table including extended columns*

The example model shown in Figure 5-5 has only four tables, and the relationships only span a single generation. If any of the three parent Dimension tables is also on the many side of a one-to-many relationship, then every column from its parent also gets included in the 'Fact Sale' virtual extended table. This expanded table is how filters applied at the top of a multilevel relationship can be applied to and affect rows in tables that are not directly related to the filtered table.

As none of the three Dimension tables has a parent table, they do not have any virtual extended columns.

Why is this important?

When building filtered calculations, it's helpful to mentally visualize tables involved in the calculation to include all extended columns from related tables as well as columns native to the base table.

At the start of any calculation, the extended table is considered clear of all filters, and potentially every row in the table is available to aggregation functions such as SUM, COUNT, MIN, and so on.

Filters get applied in layers that resolve down to a final set of rows visible to aggregation functions used in your calculations. Filters are often accumulative, but not always, and existing filters on columns can be overwritten by special DAX functions.

Unblocking plays a vital role in DAX filtering that enables calculations such as period comparisons and parent/child ratios. If filters were only accumulative and could not get unblocked, these types of calculations would be difficult to create.

Unblocking refers to the concept where a filter previously applied to a specific column in an earlier layer is removed and no longer has any effect on the column.

Layers of filtering

Filters get applied in a series of layers. To demonstrate this, consider the calculated measure in Listing 5-9. This simple calculated measure returns a count of rows from the 'Fact Sale' table that happens to include 228,265 rows. If no filters are applied, every row in the 'Fact Sale' table is visible to the COUNTROWS aggregation function, and the measure returns 228,265 every time. The reality is every time this measure is used to generate a value for a visual on the report canvas, slightly different filtering rules influence each instance.

The calculated measure in Listing 5-9 gets added to a Matrix visual and gets shown in Figure 5-7. The Matrix visual shows 20 separate instances of this calculated measure. There are five rows relating to Calendar Year that include a total row. There are four columns relating to the Buying Group that also includes a column for a total.

None of the values shown in the matrix is 228,265 (the count of rows in the 'Fact Sale' table), so each instance is affected by filtering of some form. There are no specific filtering instructions in the DAX for this calculated measure, and none of the values shown in the Matrix is the same despite each cell using the same DAX code.

The reason why a different value gets shown in every cell is that all 20 logical executions of the [Measure 5-9] calculated measure get to run in a filter context isolated from the other cells. Each context starts clear of filters, but then filters are applied to columns of the extended 'Fact Sale' table.

Listing 5-9. A simple calculated measure that uses the COUNTROWS function

```
Measure 5-9 = COUNTROWS('Fact Sale')
```

Sales Territory	Calendar Year	N/A	Tailspin Toys	Wingtip Toys	Total
2013		2,889	3,075	3,180	**9,144**
2014		3,107	3,272	3,407	**9,786**
2015		3,590	3,160	3,778	**10,528**
2016		1,468	1,357	1,480	**4,305**
Total		**11,054**	**10,864**	**11,845**	**33,763**

Slicer options: ☐ External, ☐ Far West, ☐ Great Lakes, ■ Mideast, ☐ N/A, ☐ New England, ☐ Plains, ☐ Rocky Mountain, ☐ Southeast, ☐ Southwest

Figure 5-7. *Shows the measure from Listing 5-9 in a Matrix visual next to a slicer over the [Sales Territory] column from 'Dimension City'*

Let's dissect filtering using the value of 3,272 from the second row of the second column.

The calculation starts with an unfiltered copy of an extended 'Fact Sale' table and eventually arrives at a value of 3,272 for this cell in the Matrix. This extended table includes all 228,265 rows from 'Fact Sale' and 60 columns (21 native columns + 39 related columns).

The first layer of filtering in this example comes from the slicer adjacent to the Matrix visual. The slicer is over the 'Dimension City'[Sales Territory] column and a single value selected of "Mideast." The column used by the slicer also exists in the extended 'Fact Sale' table; so a single-column table with, in this case, a single row gets added to the filter context for the calculation. If the slicer happened to have other boxes ticked, the filter table would have more rows.

This first external layer of filtering for this example gets shown in Figure 5-8 at step 1.

Capturing and studying the DAX query plan for the full query generated by Power BI for the Matrix visual helps explain the different meanings of external and internal filters.

The second and third layers of filtering come from the Matrix visual. The row header uses values from the 'Dimension Date'[Calendar Year] column, which exists in the extended copy of 'Fact Sale.'

The second row of the Matrix visual has a single value of 2014, so a single-column, single-row table is added to the filter context for this calculation and gets represented in Figure 5-8 at step 2.

The second column of the Matrix visual uses the 'Dimension City'[Buying Group] column which also exists in the extended 'Fact Sale' table. The column has a single value of "Tailspin Toys," so a single-column, single-row table is added to the filter context for this instance of the calculation and gets represented in Figure 5-8 at step 3.

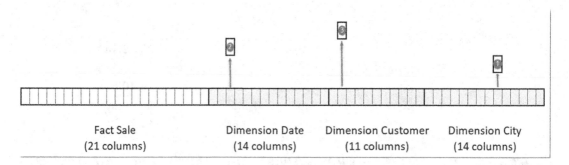

Figure 5-8. *Shows the extended 'Fact Sale' table with three columns filtered by tables of data*

The three layers of filtering added to the filter context are all external to the code of the DAX calculation and are therefore accumulative. This example contains no specific filtering directives inside the DAX code, so the COUNTROWS function is only affected by these three layers.

The 'Fact Sale' extended table is first filtered down from 228,265 to 33,763 rows by the filter table over the *Sales Territory* column. This extended table is then reduced to 3,272 rows by filter tables sitting over the *Calendar Year* and *Buying Group* columns. It's at this point the COUNTROWS aggregation function can begin its work using the extended table, and once it completes, it returns a value of 3,272 for the cell.

The same steps get logically repeated for all other 19 cells in the Matrix visual that use this calculated measure. The difference between each calculation is the content in the filter tables that get added to the filter context. These filter tables include rows using values based on the appropriate column or row header, and in the case of the totals, fewer filter tables get added to the filter context.

Unblocking filter tables

A common requirement in Power BI reports is to create calculations that need to access data in rows previously filtered out by axis values.

Consider a period comparison measure that compares a specific value for a given period with a value from a different period. Figure 5-7 shows a Matrix of values using the simple calculation in Listing 5-9. The Matrix has a value of 3,272 on the second row of the second column which represents the number of rows in the 'Fact Sale' table after filter tables are applied to three columns as shown in Figure 5-7.

A method is required to remove filtering that prevents visibility of rows relating to the Calendar Year of 2013 to help understand the percentage change between the value of 3,271 for Calendar Year 2014 and the value of 3,075 for the Calendar Year of 2013.

DAX provides a particular set of filter functions that can undo a filter on a column applied earlier in the same query. These unblocking filters are

- ALL

- ALLEXCEPT

- ALLSELECTED

When any of these functions get used in DAX calculations, they overwrite any previous filter on any column included in the table expression.

To see an unblocking function in action, the DAX calculated measure from Listing 5-9 is modified in Listing 5-10 to now include an ALL function. The *MyFilterTable* variable is assigned a table expression built using the ALL function. The ALL function could have been used as the second parameter in the CALCULATE function without the use of a variable, but I like this pattern as it highlights the fact that the ALL function is a table of rows and columns. In this example, the table expression assigned to the *MyFilterTable* variable has a single column and four rows.

Note   Using ALL directly in CALCULATE is different from using ALL indirectly through a variable.

Listing 5-10. A simple calculated measure that uses the COUNTROWS function

```
Measure 5-10 =
VAR MyFilterTable = ALL('Dimension Date'[Calendar Year])
RETURN
    CALCULATE(
        COUNTROWS('Fact Sale'),
        MyFilterTable
    )
```

Figure 5-9 shows the extended 'Fact Sale' table with three layers of filtering applied externally to the DAX calculation at steps 1, 2, and 3. The figure also shows a new fourth block that sits above one of the earlier filters. The block at step 4 represents the table expression stored in the *MyFilterTable* variable and has the effect of overwriting the table filter applied earlier in the query to the 'Dimension Date'[Calendar Year] column from the row header of the Matrix visual. This unblocking only affects the scope of this instance of the calculated measure.

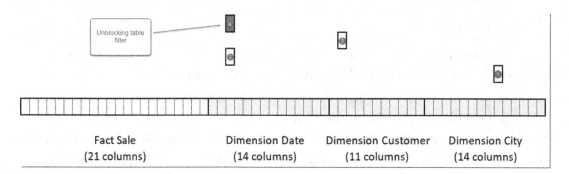

Figure 5-9. *Shows the extended 'Fact Sale' table with three columns filtered by tables of data*

The Matrix visual shown in Figure 5-10 shows the effect the updated calculation has on the values. The filter tables over the *Sales Territory* and *Buying Group* columns are unaffected. However, the same number now gets repeated in every column. The updated result reflects the *Calendar Year* column is now 100% unfiltered. The number shown in each cell is now unaffected by the *Calendar Year* column.

Calendar Year	N/A	Tailspin Toys	Wingtip Toys	Total
2013	11,054	10,864	11,845	**33,763**
2014	11,054	10,864	11,845	**33,763**
2015	11,054	10,864	11,845	**33,763**
2016	11,054	10,864	11,845	**33,763**
Total	**11,054**	**10,864**	**11,845**	**33,763**

Sales Territory: ☐ External, ☐ Far West, ☐ Great Lakes, ■ Mideast, ☐ N/A, ☐ New England, ☐ Plains, ☐ Rocky Mountain, ☐ Southeast, ☐ Southwest

Figure 5-10. *Shows the extended 'Fact Sale' table with three columns filtered by tables of data*

In this example, the ALL function targets a single column. The ALL function can also be applied to an entire table.

Listing 5-11. Shows the measure from Listing 5-10, but the ALL function gets applied to the entire table, rather than a specific column

```
Measure 5-11 =
VAR MyFilterTable = ALL('Dimension Date')
RETURN
    CALCULATE(
        COUNTROWS('Fact Sale'),
        MyFilterTable
    )
```

Figure 5-11 shows the unblocking table is now a 1,461-row table with 14 columns. Previously this was a four-row table with a single column. The result doesn't change any of the values shown in Figure 5-10 as none of the additional 13 columns from the 'Dimension Date' table in the extended 'Fact Sale' table had existing filtering to unblock.

The example in Listing 5-11 demonstrates a less efficient way to unblock columns in the filter context.

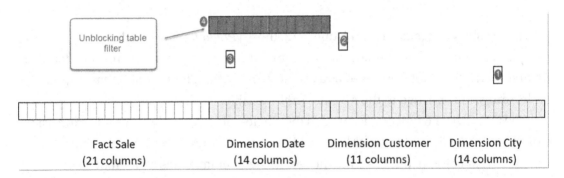

| Fact Sale | Dimension Date | Dimension Customer | Dimension City |
| (21 columns) | (14 columns) | (11 columns) | (14 columns) |

Figure 5-11. *Shows the ALL function unblocking every column from the 'Dimension Date' table*

Getting back to the original requirement to have a measure show a percentage difference between a current year and previous year. The ALL function gets used in the calculated measure in Listing 5-11 to achieve the requirement partially. To fully achieve it, an additional filter needs to be applied over the Calendar Year column *after unblocking* that filters rows to match the previous year.

This additional filter gets added to the updated calculated measure shown in Listing 5-12.

Listing 5-12. Shows the code in Listing 5-11, filtered by Calendar Year

```
Measure 5-12 =
VAR MyPreviousYear =
    SELECTEDVALUE('Dimension Date'[Calendar Year]) - 1
VAR MyFilterTable =
    FILTER(
        ALL('Dimension Date'[Calendar Year]),
        [Calendar Year]=MyPreviousYear
        )
RETURN
    CALCULATE(
    COUNTROWS('Fact Sale') ,
    MyFilterTable
    )
```

In the calculated measure in Listing 5-12, the *MyPreviousYear* variable gets assigned an integer value that subtracts 1 from the Calendar Year in the row header. The *MyPreviousYear* variable gets used as a predicate in the FILTER function that outputs a table expression assigned to the *MyFilterTable* variable.

The CALCULATE function then adds the single-column, single-row table expression stored in the *MyFilterTable* variable to its filter context that reintroduces filtering to the Calendar Year column, but this time the row in the filter table has a different value.

Figure 5-12 shows the updated filter context with the new table sitting at step 5. The first three filter tables are all external to the calculated measure. The filter table at 4 is the unblocking filter table that overwrites the filter table at 2. The new filter table at 5 reintroduces some filtering to the Calendar Year column, but using different conditions to the original filter on that column at step 2.

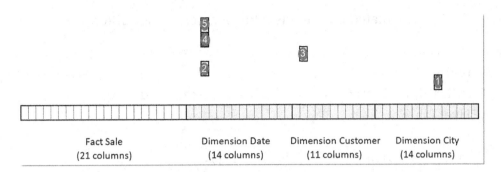

Figure 5-12. *Updated filter context with a new table filter at step 5 over the Calendar Year column in the 'Dimension Date' table*

The updated result gets shown in Figure 5-13 with each cell now showing the number of rows for the previous year.

Calendar Year	N/A	Tailspin Toys	Wingtip Toys	**Total**
2013				
2014	2,889	3,075	3,180	**9,144**
2015	3,107	3,272	3,407	**9,786**
2016	3,590	3,160	3,778	**10,528**
Total				

Sales Territ... ⌄
☐ External
☐ Far West
☐ Great Lakes
■ Mideast
☐ N/A
☐ New England
☐ Plains
☐ Rocky Mountain
☐ Southeast
☐ Southwest

Figure 5-13. *Updated Matrix visual showing the calculated measure from Listing 5-12*

The final step to meet the original requirement to show a percentage value that reflects the change in the metric from the previous year is to capture both the current and previous values in the same calculation and divide the delta value by the preferred value.

The updated measure gets shown in Listing 5-13.

Listing 5-13. A calculated measure is returning percentage difference

```
Measure 5-13 =
VAR MyPreviousYear =
    SELECTEDVALUE('Dimension Date'[Calendar Year]) - 1
VAR MyFilterTable =
    FILTER(
        ALL('Dimension Date'[Calendar Year]),
        [Calendar Year]=MyPreviousYear
        )
VAR MyPreviousValue =
    CALCULATE(
        COUNTROWS('Fact Sale') ,
        MyFilterTable
        )
VAR MyCurrentValue = COUNTROWS('Fact Sale')
RETURN
    DIVIDE(
        MyCurrentValue - MyPreviousValue,
        MyCurrentValue
    )
```

The enhancements made to the final calculated measure are to assign what was previously in the RETURN statement to a variable called MyPreviousValue. A new variable called MyCurrentValue stores the number of rows using a filter context that includes the three externally driven filter tables over *Calendar Year, Sales Territory,* and *Buying Group*. The filter context for the COUNTROWS function assigned to the *MyCurrentValue* variable is separate from the filter context used by the CALCULATE statement that assigns a table expression to the MyPreviousValue variable.

The RETURN statement derives a delta number that is divided over one of the earlier numbers to create a value that when formatted as a percentage shows how DAX filters can be manipulated to achieve a result.

Caveat

The calculated measure in Listing 5-13 is not a recommendation on the best practice for generating a period comparison measure. There are plenty of better alternatives that make use of Time Intelligence functions such as the PREVIOUSYEAR function. Functions like PREVIOUSYEAR and PARALLELPERIOD take care of the filter context work internally resulting in code that is less verbose. Listing 5-14 shows an alternative method that uses one of the Time Intelligence functions.

Listing 5-14. An alternative calculated measure that uses the PREVIOUSYEAR Time Intelligence function

```
Measure 5-14 =
VAR MyPreviousValue =
    CALCULATE(
        COUNTROWS('Fact Sale'),
        PREVIOUSYEAR('Dimension Date'[Date])
        )
VAR myCurrentValue =
    COUNTROWS('Fact Sale')
RETURN
    DIVIDE(
        myCurrentValue - myPreviousValue,
        myCurrentValue
    )
```

In Listing 5-14, the PREVIOUSYEAR function returns a single-column table expression with 365 rows (or 366 for leap years) that contains the same unblocking signal the ALL function used and overwrites any filter tables in the filter context added from sources outside the DAX calculated measure.

ALLEXCEPT

A companion function for the ALL function is ALLEXCEPT. This function has similar behavior to the ALL function when added to a filter context. The ALLEXCEPT function overwrites (unblocks) any column that may already have a filter that needs to get removed.

The difference between the ALL and ALLEXCEPT functions is syntax and how you specify which columns need to have filters overwritten.

Consider a table called 'Table1' that has five columns. The columns are called A, B, C, D, and E. Table 5-1 shows the effect and table expression for the various syntax options using the two functions.

Table 5-1. *Effect of different uses of ALL and ALLEXCEPT functions*

Syntax	Effect	Table Expression
ALL('Table1')	Overwrites filters on columns A, B, C, D, & E	Returns table expression with every distinct combination of columns A, B, C, D, & E
ALL('Table1' , 'Table1'[B], 'Table1'[D])	Overwrites filters on columns B & D	Returns table expression with every distinct combination of columns B & D
ALLEXCEPT ('Table1' , 'Table1'[B], 'Table1'[D])	Overwrites filters on columns A, C, & E	Returns table expression with every distinct combination of columns A, C, & E

A reason you might use ALLEXCEPT in place of ALL is with a table with a large number of columns that requires filters to get cleared from most but not all columns. In this case, the ALL function would work, but requires every column to be passed as a parameter that needs to have filtering unblocked.

If a table has 100 columns and you need to clear filters from 98 of those, the ALLEXCEPT function only requires the <table> parameter along with the two columns not to get unblocked. All other 98 columns would get unblocked. If the ALL function gets used, the base table along with all 98 columns would need to get added as parameters to the function. The readability of ALLEXCEPT, in this case, is much better than using the ALL function.

Under the covers, the ALL and ALLEXCEPT functions do the same thing. The only difference is how columns are selected to have previous filters cleared. It would not make sense to use both functions in the same filter context.

ALLSELECTED

The ALL and ALLEXCEPT functions overwrite (unblock) table filters previously added to a filter context regardless of how they get added to the context. Table filters can be added to a filter context either directly by the visual (the current query) or indirectly, in the form of a slicer selection external to the query.

The ALLSELECTED function provides flexibility to remove filters that have been added by a column appearing in a row or column header, or by the axis of a visual, but to keep external filters set by items such as slicers, report,

Like the ALL function, the syntax allows you to specify an entire table or target specific columns of a table to have filter context overwritten. The ALLSELECT function also provides functionality.

Note   When using ALL, ALLEXCEPT, or ALLSELECTED to overwrite a column filter, you also need to include the 'Sort by Column' if the target column happens to be sorted by another column.

A good use case for the ALLSELECTED function is with any measure that requires the total of a parent in a visual. An example of this is a ratio-type calculation like the measure shown in Listing 5-15 that needs to clear some filters to determine the overall total for the year.

Listing 5-15. An example of a parent/child ratio calculation

```
Measure 5-15 ALLSELECTED=
VAR myYearTotal =
    CALCULATE(
        COUNTROWS('Fact Sale'),
            ALLSELECTED('Dimension Customer'[Buying Group])
        )
VAR myMonthTotal = COUNTROWS('Fact Sale')
RETURN
    DIVIDE(
        myMonthTotal,
        myYearTotal
    )
```

In the example in Listing 5-15, the *myYearTotal* variable is assigned the output of a CALCULATE that makes use of the ALLSELECTED function. This statement only unblocks the columns passed as parameters to the ALLSELECTED function, for filters applied from within the visual, such as from a row or column. The ALLSELECT function does not unblock filtering coming from visuals or slicers external to the visual using the measure.

The value assigned to the myMonthTotal retains all filtering and is used in the RETURN statement to provide a ratio. The ideal formatting to use for this measure is as a percentage.

Figure 5-14 shows the percentage change calculation from Listing 5-15 along with a near-identical copy of the calculation that uses the ALL function instead of ALLSELECTED.

In step 1 shown in the top-left corner of Figure 5-14, a slicer has two out of three items selected over the [Buying Group] column. The same column is used inside the visual in the columns field at step 2.

The values shown at step 3 represent the calculated measure in Listing 5-15 when added to the matrix. The top values in the top row (33.63%, 34.78%, and a total of 68.41%) are using a version of the calculation that uses the ALL function. This version unblocks the filtering driven by the slicer at step 1, and these figures represent a breakdown over the entire [Buying Group].

The values in the second row of step 3 (49.16%, 50.84%, and 100.00%) use the version as shown in Listing 5-15 that uses the ALLSELECTED function. This version only unblocks the filtering coming from the [Buying Group] in the column header at step 2. The filters applied to the [Buying Group] column by the slicer at step 1 do not get cleared, so the values shown represent the percentage of the total, but only for the items selected by the filter.

Depending on your requirement, the solution may use either ALL or ALLSELECTED. This example is designed to demonstrate the difference between the two functions.

Buying Group ☐ N/A ① ■ Tailspin Toys ■ Wingtip Toys	Calendar Year ②	Tailspin Toys	Wingtip Toys	Total
	2013			
	Measure 5-15 ALL ③	33.63%	34.78%	**68.41%**
	Measure 5-15 ALLSELECTED	49.16%	50.84%	**100.00%**
Sales Territory	2014			
☐ External	Measure 5-15 ALL	33.44%	34.82%	**68.25%**
☐ Far West	Measure 5-15 ALLSELECTED	48.99%	51.01%	**100.00%**
☐ Great Lakes	2015			
■ Mideast	Measure 5-15 ALL	30.02%	35.89%	**65.90%**
☐ N/A	Measure 5-15 ALLSELECTED	45.55%	54.45%	**100.00%**
☐ New England	2016			
☐ Plains	Measure 5-15 ALL	31.52%	34.38%	**65.90%**
☐ Rocky Mountain	Measure 5-15 ALLSELECTED	47.83%	52.17%	**100.00%**
☐ Southeast	**Measure 5-15 ALL**	**32.18%**	**35.08%**	**67.26%**
☐ Southwest	**Measure 5-15 ALLSELECTED**	**47.84%**	**52.16%**	**100.00%**

Figure 5-14. Shows the matrix using two similar calculated measures to highlight the difference between ALL and ALLSELECTED functions

Summary

In this chapter, we have explored filtering in more detail. We have covered simply Boolean-based filtering as well as shown how tables get used as filters. Table-based filters provide set-based filtering that can be simple, powerful, and effective.

We have looked at extended columns and lineage and the role they play in the overall filtering logic. Understanding the part lineage plays in table-based filtering to define which columns get filtered and which aren't is helpful with more detail in Chapter 3, "DAX Lineage."

Table-based filters can apply filters to multiple columns at the same time, but take care to understand the OR/AND nature when applying filters in this way.

Finally, we covered functions that assist in cleaning/unblocking columns previously filtered for calculations and how these can be used to provide flexibility to solve more complex scenarios.

For additional reading, I recommend anything by Jeffrey Wang (the father of DAX) who has more information in his pbidax or mdxdax blog sites.

CHAPTER 6

Iterators

Introduction

A workhorse set of functions in the DAX language are the iterator functions. These can sometimes be known as X functions because most of these functions have an X as the last character in the function name. These functions are referred to as iterators because their fundamental purpose is to execute a DAX expression inside a series of loops. Iterator functions *iterate* over a table stopping at each row to process a DAX calculation, hence the name iterators.

Note Iterators provide functionality similar to FOREACH- or FOR IN-type functions in other languages in that they execute code inside a series of loops.

All iterator functions share the same initial parameter signature with the first parameter always needing to be a table, while the second parameter is always an expression. Third and subsequent parameters differ from function to function.

```
FUNCTION ( <table> , <expression> )
```

The first parameter <table> can either be a physical table or a table expression return value from a nested function or previously assigned to a variable earlier in the calculation.

The second parameter <expression> tells the iterator function what calculation gets processed during each iteration. The <expression> can be as simple as a hardcoded value or, more commonly, a DAX calculation that uses data from the data model.

Most iterator functions accept only these first two parameters, so there is a high degree of consistency in how they behave. However, RANKX and CONCATENATEX allow additional optional parameters to be passed to help provide more specific details on how these functions should operate.

© Philip Seamark, Thomas Martens 2019
P. Seamark and T. Martens, *Pro DAX with Power BI*, https://doi.org/10.1007/978-1-4842-4897-3_6

The following is a list of *some* of the iterator functions in DAX:

CONCATENATEX

AVERAGEX

COUNTAX

COUNTX

GEOMEANX

MAXX

MEDIANX

MINX

PRODUCTX

RANKX

SUMX

PercentileX.Exc

FILTER

GENERATE

GENERATEALL

Note   The FILTER function is an iterator; despite not having an X at the end of the function name, it behaves the same way as other iterators.

Looping flow control

When an iterator function is used in a DAX calculation, internally, it executes a series of loops. The number of rows in the <table> parameter determines the number of loops. If a <table> happens to have seven rows, the iterator function iterates precisely seven times.

Once the number of iterations is known, and looping is underway, the iterator function evaluates the <expression> specified as the second parameter at every iteration. This expression produces a single value applicable to the loop taking into account any relevant filter context. The <expression> can be as simple as a column name, but can also be a complex calculation involving many lines of DAX code, so long as it produces a single value as an output for the loop. The iterator function keeps track of every value generated by each iteration.

During the iteration process, the iterator function keeps track of the output value generated by each loop. If the SUMX iterator function gets used, the value generated by each loop would be tallied up for a total. If the AVERAGEX iterator function gets used, the iterator function produces an average of the values generated by each loop.

The ability to perform two passes over data is an essential distinction between the different functions such as AVERAGE and AVERAGEX and often the reason why you might choose to use an iterator function over the non-iterator version of the same function. With the iterator version, there is the ability to process data into the right format before performing the final average.

It's important to remember the iterator performs as many loops as there are rows in the <table> used as the first parameter. The number of iterations may still mean the <expression> performs a calculation over many more rows. Do not confuse the number of iterations between the rows considered by the <expression> and the number of rows in the <table>.

Basic form

The most basic form of an iterator gets shown in Listing 6-1. This *calculated measure* example uses the SUMX iterator to perform exactly six loops. The number of loops gets determined by the number of rows in the table expression used as the first argument.

A variable named *MyForEach* is assigned a table expression containing exactly six rows, which is the output from the GENERATESERIES function. In this example, the GENERATESERIES function outputs a single-column, six-row table. The *MyForEach* variable then gets used as the first argument of the SUMX iterator function in the RETURN statement.

Listing 6-1. A rudimentary example of an iterator function used in a calculated measure

```
SUMX Example =
VAR
    MyForEach = GENERATESERIES(1,6)
RETURN
    SUMX( MyForEach, 2 )
```

Each iteration of the SUMX function in Listing 6-1 evaluates an <expression> which in this case gets hardcoded to an integer value of 2. In this example, the output of the calculated measure is always 12. The SUMX function performs a SUM over the output of each iteration:

$$2 + 2 + 2 + 2 + 2 + 2 = 12$$

A tweak to the calculation in Listing 6-1 is to adjust the <expression> so that it uses data from the model instead of a hardcoded value. In Listing 6-2, the <expression> is extended to now use values from the table expression stored in the *MyForEach* variable.

Note   Data used in the <expression> parameter does not need to be from or have a relationship with the <table> used as the first parameter.

Listing 6-2. SUMX iterator function in a calculated measure

```
SUMX Example =
VAR
    MyForEach = GENERATESERIES(1,6)
RETURN
    SUMX( MyForEach , [Value] * 2 )
```

The updated calculation in Listing 6-2 still performs exactly six iterations; however, in this case, the <expression> for each loop returns a different output each time. The [Value] reference in the <expression> is a number between 1 and 6 depending on the iteration, which is then multiplied by 2.

The variable name *MyForEach* used here is to highlight the intent of the example and is not part of the iteration logic.

The <expression> calculated for the first iteration generates a value of 2, while the second iteration generates a value of 4, and so on. The final value returned by the SUMX function is 42.

$$2 + 4 + 6 + 8 + 10 + 12 = 42$$

This result also happens to be the same answer to the ultimate question of life, the universe, and everything. So at least you now know this famous problem can be solved in DAX using an iterator function. ☺

The example in Listing 6-2 also shows how a <expression> passed to the iterator function can be more complex than an operation over a single column.

Common use case

A common use case for an iterator function is to provide an accurate total over data that requires calculations to get executed along the way. Listing 6-3 shows a calculation where the iterator output is the [Quantity] column multiplied by the [Price] column.

Consider the following six-row dataset in Table 6-1 along with a requirement to show a row total amount per year as well as an accurate overall value for the total. The data in Table 6-1 gets used by the calculated measure in Listing 6-3.

Table 6-1. *A simple dataset to demonstrate SUM vs. SUMX*

Year	Quantity	Price
2014	2	$1.50
2014	4	$1.60
2015	6	$1.70
2015	8	$1.80
2016	10	$2.10
2016	12	$2.20

A calculated measure that satisfies the requirement uses an iterator and gets shown in Listing 6-3.

Listing 6-3. Calculated measure using the SUMX iterator function and the data from Table 6-1

```
Total using SUMX =
    SUMX(
        'Table 6-1',
        'Table 6-1'[Price] * 'Table 6-1'[Quantity]
        )
```

An additional calculated measure called [Total using SUM] gets added to the data model and gets shown in Listing 6-4. This calculated measure is helpful to show the difference between iterator and non-iterator functions, particularly when both get used in the same table or Matrix visual in Power BI.

Listing 6-4. Calculated measure using the SUM function and the data from Table 6-1

```
Total using SUM =
    SUM('Table 6-1'[Price]) * SUM('Table 6-1'[Quantity])
```

When both calculated measures are added to a table visual, the result is as shown in Figure 6-1.

Year	Price	Quantity	Total using SUM	Total using SUMX
2014	$3.10	6	$18.60	$9.40
2015	$3.50	14	$49.00	$24.60
2016	$4.30	22	$94.60	$47.40
Total	**$10.90**	**42**	**$457.80**	**$81.40**

Figure 6-1. *Shows a visual using calculations from Listing 6-3 and Listing 6-4*

Consider the first row of the table visual. The value of $18.60 in the fourth column uses the [Total using SUM] calculated measure from Listing 6-4 that does *not* iterate. It performs a single calculation over the SUM of Price which is then multiplied by the SUM of Quantity.

The SUM function adds the two values for 2014 together in the [Quantity] column to arrive at 6 and then adds the two [Price] values together of $1.50 and $1.60 before finally multiplying these two values for a total of $18.60.

$$6 \text{ x } \$3.10 = \$18.60$$

This value of $18.60 is what gets shown in the top row of the [Total using SUM] column. While $3.10 multiplied by 6 does equal $18.60, this is not the result we are after.

The [Total using SUMX] column on the right-hand side is a calculated measure that uses an iterator function to perform the same multiplication between price and quantity, but the difference is when the multiplication happens for each row.

There are two rows in the underlying dataset for the year 2014, so to produce a value of $9.40 for 2014, the iterator first calculates 2 x $1.50 to get a value of $3.00 for the first of the two iterations.

Note   SUM is the same as SUMX under the covers. SUM is *syntactic sugar* for cases where the intent is to work with a single column and not as flexible as SUMX.

The second iteration multiplies 4 x $1.60 to produce a value of $6.40. Once all iterations are complete, and the output of each expression gets established, the iterator function performs a final calculation over the values produced by each iteration. In this case, the iterator function is SUMX, so the values get added to produce a total of $9.40.

$$(2 \text{ x } \$1.50) + (4 \text{ x } \$1.60) = \$9.40$$

The difference between using an iterator function and a non-iterator function, in this case, is usually more apparent in values representing totals. The Total row in Figure 6-1 shows a value of $457.80 in the [Total using SUM] column, which is the result of a single pass calculation of the total [Price] multiplied by the total of all values of the [Quantity] column which results in a value of $457.80.

$$42 \text{ x } \$10.90 = \$457.80$$

While this calculation returns quickly, it is not the figure we are after.

> **Note**   In this case, a value of $10.90 is visible in the Total row of the Price column. The value of $10.90 was not hidden to help demonstrate the difference between the two measures. Usually, this value should be hidden to prevent confusion.

Average of an average

Another typical scenario that takes advantage of iterators is a requirement to perform an average over another average. Without getting into a debate on the statistical merit of such a requirement, consider how you might go about finding out the average annual price using the same dataset from Table 6-1.

A particular challenge with this data is to determine the average price for any given year; the calculation needs to take into account the quantity sold at each price point. For 2014 there are more items sold at the $1.60 price than at $1.50. The average price value for 2014 should not merely be the midpoint between the two price values; it should reflect a weighting based on the values in the Quantity column.

Table 6-2. *The 2014 values from Table 6-1 with expanded rows to show the weighting*

Year	Quantity	Price
2014	1	$1.50
2014	1	$1.50
2014	1	$1.60
2014	1	$1.60
2014	1	$1.60
2014	1	$1.60
TOTAL	6	$9.40

Based on the *expanded* values in Table 6-2, the average price for sales in 2014 was $1.5667.

$$\$9.40/6 = \$1.5667$$

Using the same logic, the average price for 2015 was $1.7571, and the average price for 2016 was $2.1545.

Table 6-3. *Average values for each year from Table 6-1 including average over the three years*

Year	Average Price
2014	$1.5667
2015	$1.7571
2016	$2.1545
AVERAGE	$1.8261

The objective is to create a calculated measure that can be added to a table visual in a Power BI report that generates the same values shown in Table 6-3.

The correct result would be difficult to achieve without the use of iterator functions. A simple SUM over the Price column combined in a calculation with a SUM over the Quantity column, as per the calculation in Listing 6-4, does not have enough fine-grained access to the data required to satisfy this requirement.

To make the final calculation easier to read and also force row context to transition, the following two calculated measures shown in Listings 6-5 and 6-6 get added to the data model.

Listing 6-5. A calculated measure producing a SUM of the values in the Price column

```
Sum of Price = SUM('Table 6-1'[Price])
```

Listing 6-6. A calculated measure producing a SUM of the values in the Quantity column

```
Sum of Quantity = SUM('Table 6-1'[Quantity])
```

Nested iterators

A handy feature of iterators is the ability to nest them, which means an outer iterator function can happily call an inner iterator. The solution suggested for this exercise uses nested iterators to help with the necessary processing to achieve its result. Listing 6-7 uses two iterators in this fashion as part of the calculation.

The outer AVERAGEX iterator performs three iterations because the table expression returned by the VALUES function has exactly three rows. The VALUES function returns a single-column table with a row for each of the distinct values in the Year column from Table 6-1.

An important thing to note here is the AVERAGEX function performs three loops and not six based on the number of rows in Table 6-1. The VALUES function aggregates the raw data from six rows down to three.

The inner SUMX iterator function executes each time the outer AVERAGEX function iterates and itself runs as many iterations as there are rows in the table used as the first argument. With this dataset, the SUMX function happens to perform two iterations for each year in context.

Listing 6-7. A calculated measure showing the average price each year

```
Total using AVERAGEX =
    AVERAGEX(
        VALUES('Table 6-1'[Year]),
        DIVIDE(
            SUMX(
                FILTER(
                    'Table 6-1',
                    'Table 6-1'[Year] =
                        EARLIEST('Table 6-1'[Year])
                ),
                [Sum of Price] * [Sum of Quantity]
            ),
            [Sum of Quantity]
        )
    )
```

Year	Total using AVERAGEX
2014	$1.5667
2015	$1.7571
2016	$2.1545
Total	**$1.8261**

Figure 6-2. *Shows the result from the calculation in Listing 6-7*

The pseudo-logic of the code shown in Listing 6-7 (with results shown in Figure 6-2), can get converted to the following step-by-step logic:

1. Run an outer loop (AVERAGEX) for each distinct year in Table 6-1. There are three iterations in total for the outer loop, one for each year (2014, 2015, and 2016).

2. For each of the three outer loops in step 1, run an inner loop using SUMX over the raw data from Table 6-1. Each inner loop gets filtered to the year of the current outer loop. This filtering results in two iterations for each inner loop of the SUMX function.

3. Each iteration of the SUMX evaluates Price x Quantity to determine an overall total for the row currently being iterated using the table from the first argument of the inner SUMX function.

4. Once the inner SUMX has completed iterating, each result is added to determine a total for the inner loop.

5. The result of the inner SUMX gets divided by the sum of Quantity for the current outer loop to determine the average sales quantity for the individual year. This division happens as part of the <expression> for the outer AVERAGEX loop.

6. An average gets computed over the output value for each iteration of the inner function which is the output of step 5. There are three results to be considered by the average.

Listing 6-8. Pseudocode is showing loops and values of the DAX function in Listing 6-7

```
{
2014 :
            {
                  { 2 x $1.50 = $3.00 }
            +
                  { 4 x $1.60 = $6.40 }
            } / 6 = $1.5667
      +
2015 :
            {
                  { 6 x $1.70 = $10.20}
            +
                  { 8 x $1.80 = $14.40 }
            } / 14 = $1.7571
      +
2016 :
            {
                  { 10 x $2.10 = $21.00 }
            +
                  { 12 x $2.20 = $26.40 }
            } / 22 = $2.1545
      } / 3 = $1.8261
```

To arrive at an outcome of $1.8261, six iterations in total get performed in Listing 6-8. There are three outer loops with each outer loop running two inner loops. These inner/outer loops demonstrate how iterator functions can provide a fine-grained access to data that allows calculations to execute *before* aggregation takes place. This example does not mean you should always use SUMX in place of SUM, or AVERAGEX in place of AVERAGE. This example is designed to show how different DAX functions are suited to different requirements, and the iterator functions can make *some* complex scenarios much easier to solve.

EARLIER and EARLIEST

When nesting iterators like the example in Listing 6-7, the inner iteration sometimes needs access to a value outside its iteration logic. Two DAX functions to help with this type of requirement are EARLIER and EARLIEST. These functions allow a calculation to access a current value that is higher in the call stack. Despite the sound of these function names, they are not designed to solve date- or time-based challenges directly, nor do they give you direct access to data in other rows in the same table – such as a value from a row above or below in the same table.

There is no corresponding LATER or LATEST function.

EARLIER

To demonstrate these functions using nested iterators, consider the following calculated measure in Listing 6-9. The intention is to build a text-based string using numbers and letters that help to highlight the flow control of nested iterators.

There are two iterators used in this calculation, an outer iterator that loops through numbers 1–4 and the inner iterator that loops through letters A, B, and C. The result is an item of text that nicely shows the order of how inner and outer loops operate.

The calculation in Listing 6-9 also includes some comment lines which begin with double-slash (//) characters.

Listing 6-9. A calculated measure showing EARLIER in a nested iterator

```
Loop Measure =
VAR outerLoop = GENERATESERIES(1,4)
VAR innerLoop = DATATABLE("Value",STRING,{{"A"},{"B"},{"C"}})
RETURN
    CONCATENATEX(
        // foreach row in the outerLoop table
        outerLoop ,
        CONCATENATEX(
            //foreach row in the innerLoop table
            innerLoop ,
            EARLIER([Value]) & "-" & [Value] ,
```

```
        //inner loop delimiter
        " , "
        ) ,
    //outerLoop Delimiter
    " / "
    )
```

```
1-A , 1-B , 1-C /
2-A , 2-B , 2-C /
3-A , 3-B , 3-C /
4-A , 4-B , 4-C
```

Figure 6-3. *Shows a visual using the output of the calculated measure from Listing 6-9*

In the calculation shown in Listing 6-9, there are two CONCATENATEX functions. The first CONCATENATEX function immediately following the RETURN statement is the outer loop. The first argument provided is the table expression stored in the *outerLoop* variable. This table expression is a single-column table with four rows. The values in this table are numbers 1–4.

The other CONCATENATEX function is nested inside the second argument of the initial CONCATENATEX and uses the table expression stored in the *innerLoop* variable to determine the number of loops for this iterator.

The outer CONCATENATEX function starts with the first row of the table expression stored in the *outerLoop* variable which has a value of 1. The inner CONCATENATEX function then kicks in and iterates through the three rows stored in the table used to drive the inner iterator.

The <expression> used to build each section of text is the following line (line 11):

```
EARLIER([Value]) & "-" & [Value] ,
```

In this code, a reference to a [Value] column appears twice. The tables used for both inner and outer iterators are single-column tables that use a column name of [Value]. The EARLIER function helps the calculation distinguish the correct column reference to use.

In this case, the [Value] column reference inside the EARLIER function refers to the table expression in the outer loop, so it returns a value from the current row of the *outerLoop* table function. The second [Value] in the code does not use the EARLIER function and refers to the [Value] column belonging to the current iteration of the *innerLoop* table being iterated over by the inner CONCATENATEX function.

The delimiter used to denote an iteration of the *innerLoop* table is a comma, while the delimiter used to highlight an iteration of the *outerLoop* table is a slash. These delimiters can be seen in the result in Figure 6-3 with commas representing the progression of the inner loop and a slash showing where an outer loop has iterated.

EARLIEST

The example used in Listing 6-9 uses two nested iterator functions. Nothing is stopping you from nesting three or more iterators in a single statement. In this event, there may be a requirement in some of the inner iterators to access a current value from one of the outer iterators, and not necessarily the immediate outer iterator or the outermost.

A second optional parameter <number> for the EARLIER function allows you to specify which outer loop in the call stack you intend to reference. If a value of 1 gets used, the loop immediately outside the current iterator gets referenced, so it is no different from leaving this parameter empty.

Note   EARLIER(<table> , 1) is the same as using EARLIER (<table>).

This optional parameter allows the iterator <expression> to access the current row context of a table used in any of the outer loops by specifying a value higher than 1.

The example shown in Figure 6-4 has four layers of nested iterators. The EARLIER function in the innermost iterator specifies a value of 2 for the second parameter. This value means the <column> parameter of the EARLIER function is a reference to the current row context of the iterator *two* layers higher in the series of nested loops. A value of 3 would reference the current row context of Iterator 1, which has the same effect as specifying EARLIEST(<column>).

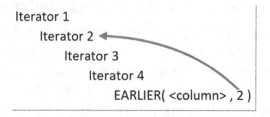

Figure 6-4. *Shows the EARLIER function at the fourth level referencing a column at the second level*

If the number specified in the optional second argument is higher than the number of nested iterators, an error gets generated.

The EARLIEST function provides a shortcut syntax to a reference to the outermost iterator, which saves the need to work out the correct number to use as a parameter for the EARLIER function.

Using the code from Listing 6-9, the following three variations shown in Listing 6-10 would produce the same result. However, if there were three or more layers of nested iterators, the variation using the EARLIEST function might produce a different result.

Listing 6-10. Three variations of EARLIER and EARLIEST that in this case return the same result

```
EARLIER([Value]) & "-" & [Value] ,

EARLIER([Value],1) & "-" & [Value] ,

EARLIEST([Value]) & "-" & [Value] ,
```

Debugging iterators

As shown in this chapter, iterator functions can be beneficial when working with complex scenarios. It's also easy to lose track of or be confused by the internal workings of what is going on with calculations using any of the various iterator functions. They output a single value, and it's not always clear if the value returned by the function gets calculated the way you intended.

Fortunately, there is a relatively easy way to debug iterator functions, and this is by using one of the iterator functions to provide a simple form of debugging. The function I use for debugging is the CONCATENATEX function.

All the iterator functions share the same parameter signature or a typical pattern for the first two parameters. The first argument is always a <table>, which can either be a physical table or a table expression. The second argument is always the <expression> to be evaluated by each iteration.

This typical pattern provides an easy way to clone the calculation you would like to debug and substitute the iterator function to debug with the CONCATENATEX function.

To demonstrate this, consider the following table of data (Table 6-4) and a requirement to create a simple calculated measure that ranks items in the Category column by the number in the Value column. The plan is to generate a calculated measure that displays a number between 1 and 5, with the highest value having a rank of 1 and the lowest value having a rank of 5.

Table 6-4. *A set of data to be used for a ranking calculation*

Category	Value
A	100
B	10
C	55
D	2
E	200

The calculated measures shown in Listing 6-11 get added to the data model. The [My Rank] calculated measure makes use of the [Sum of Value] measure.

Listing 6-11. Two calculated measures added to the data model. [Sum of Value] performs a simple sum, and [My Rank] is the measure to be debugged

```
Sum of Value = SUM('Table 6-4'[Value])

My Rank =
    RANKX(
        'Table 6-4',
        [Sum of Value]
        )
```

At first glance, the DAX used for this calculation using the RANKX iterator looks like it produces a value between 1 and 5 when used in a visual that also contains the [Category] column. However, the actual result shown in Figure 6-5 shows the [My Rank] measure repeating the value of 1 in every row.

Category	Sum of Value	My Rank
A	100	1
B	10	1
C	55	1
D	2	1
E	200	1
Total	**367**	**1**

Figure 6-5. *The [My Rank] measure is showing a value of 1 in every row*

The [My Rank] measure is generating the same value over and over, including the Total row at the bottom.

To understand why the calculation behaves this way, one method to debug is to clone the measure and substitute the RANKX iterator function with the CONCATENATEX function as shown in Listing 6-12.

Listing 6-12. A calculated measure using CONCATENATEX to debug

```
My Debug Rank =
    CONCATENATEX(
        'Table 6-4',
        [Sum of Value],
        -- Formatting Parameters --
        ", "
    )
```

Figure 6-6 shows the table visual from Figure 6-5 once the new [My Debug Rank] calculated measure from Listing 6-12 gets added. A key thing to note in the new column is in each of the top five rows, the output only shows a single value. The single output is because the [My Rank] and [My Debug Rank] calculations take into account filtering that gets applied from the [Category] column.

The top row of the table visual in Figure 6-6 has a value of "A" in the [Category] column which becomes an implicit filter over calculations in the same row that also use 'Table 6-4'. This filtering causes the table used by the RANKX and CONCATENATEX functions to get reduced from the original five rows down to a single row for the [My Rank] and [My Debug Rank] measures.

Therefore, the RANKX function only gets to consider a single value, which is why it outputs a value of 1 as a result. The CONCATENATEX function helps to clarify this by highlighting the combined effect of implicit and explicit filters at play and suggests the original calculation should get modified with code that clears the implicit filter coming from the [Category] column selection.

Category	Sum of Value	My Rank	My Debug Rank
A	100	1	100
B	10	1	10
C	55	1	55
D	2	1	2
E	200	1	200
Total	**367**	**1**	**100, 10, 55, 2, 200**

Figure 6-6. *Table visual showing the [My Debug Rank] measure*

So, armed with the helpful information provided by the [My Debug Rank] debug measure, an updated version of the calculation to address this gets shown in Listing 6-13.

Listing 6-13. An updated version of the debug measure from Listing 6-12

```
My Debug Rank =
    CONCATENATEX(
        ALL('Table 6-4'[Category]),
        [Sum of Value],
        -- Formatting Parameters --
        " , " ,
        -- Order by  --
        'Table 6-4'[Sum of Value] ,
        ASC
        )
```

The main change is that the ALL function gets wrapped around the table passed as the first parameter. This change specifically clears any internal or external filtering that exists over the [Category] column. Additional parameters have been added to the CONCATENATEX function to apply a sort order to the output of the concatenation. The additional parameters are purely for cosmetic reasons and designed to highlight how the RANKX function behaves visually.

Figure 6-7 shows the table visual using the updated [My Debug Rank] measure. Each row in the right-hand column now shows that five iterations take place on each of them, which is better than the single value shown previously.

The update to the debug measure looks good, so the same change can get applied to the [My Rank] measure which should resolve the issue.

Category	Sum of Value	My Rank	My Debug Rank
A	100	1	200, 100, 55, 10, 2
B	10	1	200, 100, 55, 10, 2
C	55	1	200, 100, 55, 10, 2
D	2	1	200, 100, 55, 10, 2
E	200	1	200, 100, 55, 10, 2
Total	**367**	**1**	**200, 100, 55, 10, 2**

Figure 6-7. *Table visual showing the updated [My Debug Rank] measure*

The updated code for the [My Rank] measure is shown in Listing 6-14.

Listing 6-14. A fixed [My Rank] calculation once debugging is complete

```
My Rank =
    RANKX(
        ALL('Table 6-4'[Category]),
        [Sum of Value]
        )
```

Once the [My Rank] calculation is modified, the updated table visual is as shown in Figure 6-8.

Category	Sum of Value	My Rank	My Debug Rank
A	100	2	200, 100, 55, 10, 2
B	10	4	200, 100, 55, 10, 2
C	55	3	200, 100, 55, 10, 2
D	2	5	200, 100, 55, 10, 2
E	200	1	200, 100, 55, 10, 2
Total	**367**	**1**	**200, 100, 55, 10, 2**

Figure 6-8. *Table visual showing the updated [My Rank] measure from Listing 6-14*

The [My Rank] column now displays the value you might expect for each [Category] based on the [Sum of Value] compared with the other categories. The default behavior of RANKX is to return a ranking of 1 for the highest value in the series and 2 for the next highest. In Figure 6-8, boxes are used in the [My Debug Rank] column to show where each [Sum of Value] sits for each specific row in the sorted series.

In the first row for Category "A," the [Sum of Value] is 100 and sits in the second position in the series when reading the values from left to right (highest to lowest). In the second row for Category "B," the [Sum of Value] of 10 sits in the fourth position using the same read order. This position reflects the number generated by the RANKX function.

More debugging

The previous section demonstrated how the CONCATENATEX function could be used to help debug a RANKX function. The same technique is equally useful for other iterator functions.

Consider the following table of data (Table 6-5) that includes four items grouped into two categories. There are five transaction rows in total in the table, and the requirement is to show a summary of the total by Category in a table visual. The total needs to be a calculation that multiplies the [Quantity] and [Value] columns at the transaction level.

Table 6-5. *A set of data to be used to help demonstrate debugging a SUMX calculation*

Category	Item	Quantity	Value
A	A1	2	100
A	A2	3	10
A	A2	2	55
B	B1	3	2
B	B2	2	200

Listing 6-15 shows a calculated measure that uses the SUMX iterator function to work out the total per Category as well as provide an accurate figure for a grand total.

Listing 6-15. A measure to calculate a total using data from Table 6-5

```
Total =
    SUMX (
        'Table6-5' ,
        'Table6-5'[Value] * 'Table6-5'[Quantity]
    )
```

When the calculated measure from Listing 6-15 gets added to a table visual along with the Category field, the result is as shown in Figure 6-9.

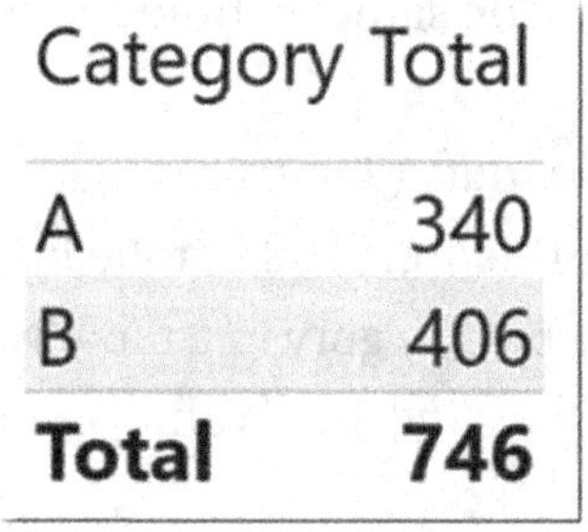

Category	Total
A	340
B	406
Total	**746**

Figure 6-9. *Table visual showing data from Table 6-5 with the measure from Listing 6-15*

The numbers shown in the right-hand column appear to be correct, but it is easy to be tripped up with numbers that look to be correct but turn out to be wrong, especially when it comes to the values in a total row at the bottom. How can we be sure the value of 746 shown in the total has been derived using the correct grain?

Alternatively, can we show the value for the total is not merely a SUM over the Value column multiplied by the SUM of the Quantity column?

The technique of substituting the SUMX function with a CONCATENATEX function helps clarify how the SUMX function arrives at each of the three values shown in Figure 6-9.

The first step is to take a copy of the calculated measure shown in Listing 6-15 and make the following adjustments to the cloned calculation.

Listing 6-16. A new measure to debug the total using data from Table 6-5

```
Total (Debug) =
    "=" &
    CONCATENATEX(
        'Table6-5' ,
        'Table6-5'[Value] * 'Table6-5'[Quantity],
        "+"
    )
```

Aside from changing the name of the original calculation, the SUMX function gets replaced by the CONCATENATEX function in Listing 6-16. The final output of the [Total (Debug)] measure is text, so a "=" gets added as a prefix, and a "+" sign gets specified as the delimiter for the CONCATENATEX function.

The new function gets added to the table visual alongside the original calculated measure, and the result is as shown in Figure 6-10.

Category	Total	Total (Debug)
A	340	=200+30+110
B	406	=6+400
Total	**746**	**=6+30+110+200+400**

Figure 6-10. *Table visual showing data from Table 6-5 with code from Listings 6-15 and 6-16*

The new [Total (Debug)] calculated measure shows a more detailed breakdown for each of the three lines about how the total figure is derived. The value of 340 in the top row for Category A is the result of adding 200, 30, and 110 together. The output of the debug measure highlights three iterations get performed even though item A2 in Category A appeared twice. The iterator still evaluates the <expression> for every line of the table used in the first parameter of the iterator.

The value of 406 for Category B is the result of adding 6 (3 x 2) and 400 (2 x 200), which indicates the iterator only performed two iterations to arrive at this number.

An essential figure in the visual is the value of 746 in the Total row. The output of the debug measure indicates that five iterations took place. These evaluations are logically separate and independent of calculations taking place in the first and second rows. The result in the total row shows the value gets calculated in parallel to other calculations in the same column.

The text-based output of the debug calculation also helps to confirm the result of 746 is not generated by summing the Value column and then multiplying it with a total over the Quantity column.

MDX SCOPE equivalent

To highlight the independent nature of the calculation for the total, the following modifications made to the calculation only affect the overall total. If you are lucky enough to have worked with Multidimensional data models, a handy feature is the SCOPE command. The SCOPE command allows the data modeler to embed specific code into measures that allow you to override logic in specific circumstances.

DAX does not have a SCOPE function, but the same effect can be achieved using the technique shown in Listing 6-17.

The calculation in Listing 6-15 is updated to show the SCOPE-like effect. The calculation of [Value] * [Quantity] gets doubled, but only if the row happens to be the Total.

Listing 6-17. An updated version of the measure in Listing 6-15

```
Total =
    SUMX (
        'Table6-5' ,
        'Table6-5'[Value]
            * 'Table6-5'[Quantity]
```

```
* IF(
    ISINSCOPE('Table6-5'[Category]), --<= TEST
    1, --<= THEN
    2  --<= ELSE
    )
)
```

The modified measure in Listing 6-17 shows an IF function added to the logic in the <expression> of the SUMX function. The function checks the filter context to see if any filtering is applied to the Category column. If the ISINSCOPE function returns TRUE, then the expression is multiplied by 1 to create no change.

However, if the ISINSCOPE function returns FALSE, the absence of filtering over the Category column implies the calculation is for the Total row, so it multiplies by 2. The result shown in Figure 6-11 assumes the modification highlighted in Listing 6-17 applied to both the [Total] and the [Total (Debug)] measures.

Category	Total	Total (Debug)
A	340	=200+30+110
B	406	=6+400
Total	**1492**	**=12+60+220+400+800**

Figure 6-11. *Table visual showing data from Table 6-5 with code from Listing 6-17*

The result shown in Figure 6-11 shows the first two lines have not changed when compared with Figure 6-10. However, the numbers displayed in the bottom row show the effect the updated code has on the output. The text in the [Total (Debug)] column highlights how the calculation is logically independent of other calculations in the same column. While the results make no sense in real terms, neither will many of the business rules you may be asked to apply in your data model. ☺

Query Plans

How do I know if a function is an iterator, and can I determine this by looking at the query plan? Chapter 13, "Query Plans," goes into much more detail on Query Plans, what they are, and how to read them. The quick answer is that it's not always clear from reading a query plan that a function is an iterator.

The DAX engine is smart enough to devise efficient plans that do not iterate, even if you have correctly used a function that you know is an iterator such as SUMX. It's possible for three different queries to yield the same physical plan, with only subtle differences in the logical plan. The logical plan may offer some clues if the underlying engine is using an iterative (row by row) approach.

Summary

This chapter has looked at iterator functions in detail. Examples have been used to show the nature of how iterator functions operate and how they differ from similar non-iterator versions, particularly around the order calculations get processed.

Iterator functions can be nested to create multiple layers of looping, and functions such as EARLIER and EARLIEST can be used to help inner loops reference current values from higher up the call stack.

The CONCATENATEX function is a useful function when it comes to debugging calculations that use other iterator functions.

All in all, they are a handy set of functions when appropriately used, and…

…you would be loopy not to use them. Ha!

CHAPTER 7

Filtering Using Measures

Introduction

Throughout this chapter, various techniques are used to demonstrate possibilities that arise when measures are not just used to calculate a numeric value but their use is extended to flag items if they meet a condition or, worse, if they do not meet a certain rule, meaning they violate a rule. These techniques are ranging from simple filtering to a more complex concept of dynamic binning. A use case for the latter will be to create a visual (to be precise a ribbon chart) that counts the number of customers whose activities can be grouped into a given bucket.

Note Throughout this section, the Power BI file "CH7 – Filtering using measures. pbix" is used. Please be aware that there is a page-level filter applied that prevents the Customer Key 0 for showing up, as this key represents the unknown customer.

On a first thought, spending a whole chapter on something that may look as simple as filtering using measures may seem to be a waste of precious space, but as often it will turn out that this is not as simple as one might guess.

This chapter is about filtering; filtering of rows is an essential part of creating measures. Creating measures always comes with two subtleties:

- The Algorithm – The complexity of the algorithm applied to filtered rows. Sometimes the underlying math is much more complex than just simply multiplying two numeric values.

- The Filtering of Rows – Sometimes determining a specific set of rows is much more complex than the algorithm that has to be applied.

© Philip Seamark, Thomas Martens 2019
P. Seamark and T. Martens, *Pro DAX with Power BI*, https://doi.org/10.1007/978-1-4842-4897-3_7

This chapter is about the latter, filtering values and/or rows and especially using measures that have to be evaluated to determine if a certain value/row will be considered or has to be omitted.

Why is special care necessary

In addition to Chapter 5, "Filtering in DAX," here some additional subtleties are covered. These subtleties have to be remembered whenever a measures will be used inside a CALCULATE, not as first parameter, the numeric expression, but instead inside the Boolean expression that finally create a filter table.

Note   This section is referencing the report page "The unexpected result."

To make these subtleties shine, let's assume that a very basic measure is defined that calculates the average quantity. This measure is shown in Listing 7-1.

Listing 7-1. Average Quantity

```
Average Quantity =
AVERAGE('Fact Sale'[Quantity])
```

The preceding measure is nothing unusual and just a starting point. For this, it's necessary to get things a little more complicated. Hence, the measure Total Quantity above average is created. This measure is shown in Listing 7-2.

Listing 7-2. Total Quantity above average

```
Total Quantity above average =
CALCULATE(
    [Total quantity]
    , 'Fact Sale'[Quantity] > [Average Quantity]
)
```

Unfortunately, this measure cannot be created as the error from Figure 7-1 is raised.

A function 'CALCULATE' has been used in a True/False expression that is used as a table filter expression. This is not allowed.

Figure 7-1. *CALCULATE cannot be used in a TRUE/FALSE expression*

This is due to some restrictions that must be considered when measures are referenced inside a filter expression. These restrictions are listed in the official documentation of the CALCULATE function. This documentation can be found here:

`https://docs.microsoft.com/en-us/dax/calculate-function-dax`

> *The following restrictions apply to Boolean expressions that are used as arguments:*
>
> - The expression cannot reference a measure.
>
> - The expression cannot use a nested CALCULATE function.
>
> - The expression cannot use any function that scans a table or returns a table, including aggregation functions.

Considering these restrictions, an approach to rewrite the measure like the one in Listing 7-3 can come to mind.

Listing 7-3. Total Quantity above average v2

```
Total Quantity above average v2 =
CALCULATE(
    [Total quantity]
    ,FILTER(
        'Fact Sale'
        , 'Fact Sale'[Quantity] > [Average Quantity]
    )
)
```

But also this measure does not work as expected even if it does not raise an error. This measure does not return any value, as shown in Figure 7-2.

Calendar Year Label	Total quantity	Average Quantity	Total Quantity above average v2
CY2013	1576744	39.63	
CY2014	1638868	39.14	
CY2015	1692502	38.15	
CY2016	759497	42.33	
Total	**5667611**	**39.37**	

Figure 7-2. *Measure Total Quantity above average v2 – no values*

Even if the reason is not obvious, the explanation is simple. FILTER is a table iterator function that creates a row context for each of the rows inside the table. As a measure implicitly performs a context transition, meaning transforming the current row context into a filter context, this new filter context is used to evaluate the condition of FILTER. This simply means that the AVERAGE is calculated for each single row; for this, the condition can never become TRUE. Each row of the table iterator will be omitted, resulting in a blank column. This can be checked if the measure Total Quantity above average v3 is dragged to the table; this measure uses equals inside the condition. It has to be noted that now the value of the Total row matches the value of the column Total Quantity.

To make the measure behave as expected, it is necessary to evaluate the measure Average Quantity outside of the table iteration. One possible solution is shown in Listing 7-4

Listing 7-4. Total Quantity above average v4

```
Total Quantity above average v4 =
var theAverage = [Average Quantity]
return
CALCULATE(
    [Total quantity]
    ,FILTER(
        'Fact Sale'
        , 'Fact Sale'[Quantity] > theAverage
    )
)
```

The preceding listing shows that the evaluation of the measure Average Quantity happens even outside the CALCULATE; the average value is stored to the variable theAverage. Inside the data model, another measure Total Quantity above average v5 is defined. This measure brings the evaluation of the average measure a little closer to the table iterator, but it's still evaluated outside the iteration. On a first glance, it seems convenient to declare variables outside of the CALCULATE, but it's a good practice to declare the variable as close as possible to the place where it is needed. This is especially true when measures are starting to become more complex.

Nevertheless, it's also considered to keep the filter table as small as possible as this makes measure evaluation much more faster, as less data has to be consumed during measure evaluation.

For this, the final measure Total Quantity above average final is defined. See Listing 7-5.

Listing 7-5. Total Quantity above average final

```
Total Quantity above average final =
CALCULATE(
    [Total quantity]
    ,
    var theAverage = [Average Quantity]
    return
    FILTER(
        ALL('Fact Sale'[Quantity])
        , 'Fact Sale'[Quantity] > theAverage
    )
)
```

The measure makes use of the DAX function ALL, one of the DAX functions that can be used to unblock existing filters from a column.

All the measures and their results are shown in Figure 7-3.

Calendar Year Label	Total quantity	Average Quantity	Total Quantity above average v2	Total Quantity above average v3	Total Quantity above average v4	Total Quantity above average v5	Total Quantity above average final
CY2013	1576744	39.63		1576744	1353783	1353783	1353783
CY2014	1638868	39.14		1638868	1401417	1401417	1401417
CY2015	1692502	38.15		1692502	1441978	1441978	1441978
CY2016	759497	42.33		759497	649400	649400	649400
Total	**5667611**	**39.37**		**5667611**	**4856658**	**4856658**	**4856658**

Figure 7-3. *All the average measures*

Tip   Place variables as close as possible to the location where they are used and keep the filter table as small as possible.

Simple filtering

This chapter will start with a very basic use case, filtering (meaning finding) customers that surpass a certain number of activities.

Note   This section is referencing the report pages "Simple Filtering" and "Simple Filtering – more complex."

Listing 7-6 shows the measurement using the DAX function DISTINCTCOUNT.

Listing 7-6. No Of Activities

```
No Of Activities =
var NoOfActivities = DISTINCTCOUNT('Fact Sale'[Sale Key])
return
IF(ISBLANK(NoOfActivities),0,NoOfActivities)
```

It is a good practice to separate the measurement and the check into two measures.

Listing 7-7 shows the measure that performs the check, meaning determines if a value meets the given criteria.

Listing 7-7. meets Activity threshold

```
meets Activity threshold =
IF([No Of Activities] > 20 , 1 , 0)
```

The measure meets Activity threshold returns 1 if the number of activities is greater than 20; otherwise, the measure will return 0.

Figure 7-4 shows how both measures are used to identify all the customers for a given month that do not satisfy the condition of the threshold measure.

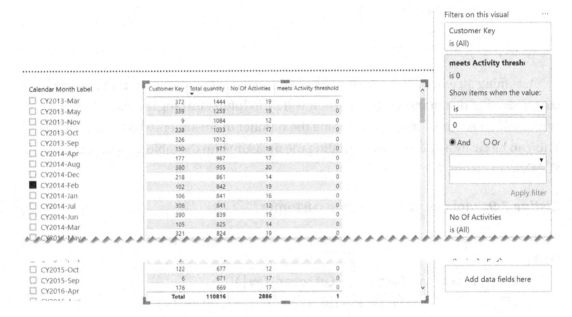

Figure 7-4. *Customers that do not satisfy the threshold*

The preceding filtering is considered simple. To make things a little more complex, the condition is changed. Now the activity threshold will be defined as the following.

A customer will be flagged if the following comes **not** true:

- If the [Total quantity] is less than 1500, the [No Of Activities] has to be less than 10.

For this, a second measure is defined; see Listing 7-8.

Listing 7-8. meets Activity threshold v2

```
meets Activity threshold v2 =
var NoOfActivities = [No Of Activities]
var TotalQuantity = [Total quantity]
return
IF(
    AND(
        TotalQuantity < 1500
        , NoOfActivities < 10
    )
    , 0 , 1 )
```

Assuming an exception reporting has to be created, where the focus is to analyze customers that do not meet a given rule, it will be helpful returning the value 1 as these values will be summed up, now returning the number of customers who violate the rule. This can be easily achieved by embedding the measure into the table iterator function SUMX as in Listing 7-9.

Listing 7-9. violates rule - counting customer

```
violates rules - counting customer =
SUMX(
    VALUES('Dimension Customer'[Customer Key])
    , IF([meets Activity threshold v2] = 0 , 1 , BLANK())
)
```

Figure 7-5 shows the application of the preceding measure. The screenshot has been taken from the report page "Simple Filtering."

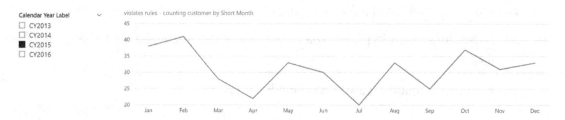

Figure 7-5. *Rule violation throughout the year*

It should be noted that the following pattern can be used to filter with measures (see Figure 7-6).

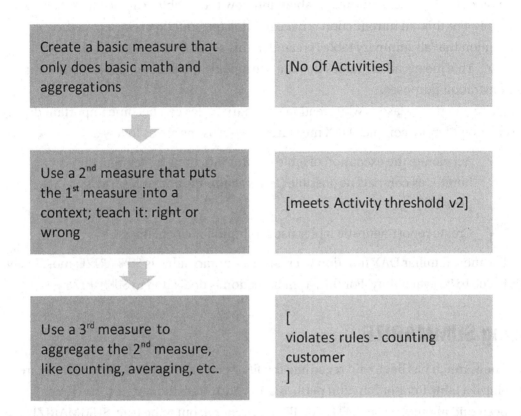

Figure 7-6. *A basic pattern*

The preceding pattern seems simple, but helps to avoid confusion if one is adhering to these steps.

Caution   Reusing measures in a way as shown in the preceding text can lead to unexpected results, as a measure automatically performs a context transition, meaning transforming an existing row context into a filter context.

Summary tables and measures

This is about using inside summary tables and how these tables can be filtered using measures. For this, an introduction is needed. This introduction is based on the assumption that all summary tables created in this section will be used later inside a measure. This means all the tables that are used inside the pbix are solely created for demonstration purposes.

There are many reasons why creating summary tables can become important during the creation of more complex DAX measures. Two reasons are as follows:

- Accelerate the execution of table iterator functions by keeping the iterator as compact as possible (as less columns and as less rows as possible).

- Create report-agnostic tables used as input for calculations.

The most familiar DAX function to create a summary table is SUMMARIZE; most likely, this is due to its availability. For this, a small section is dedicated to SUMMARIZE.

Using SUMMARIZE

Until now, much has been said regarding the filter context. Especially, discussions are ongoing if a table iterator function performs a context transition or not. And much has been said about the use of SUMMARIZE in general, but to be true, SUMMARIZE still has its merits. This is mainly because of newer functions like GROUPBY or SUMMARIZECOLUMNS. SUMMARIZECOLUMNS is not available if a measure has to be calculated, and even if this is a book about DAX, it is assumed that most of the readers are not involved in query authoring.

Note   This section is referencing the report page "Summary tables and measures."

There are some reasons why SUMMARIZE should be used only to summarize data, creating a summary table. And to add Calculated Columns, it is strictly recommended to add these columns using the following pattern:

```
ADDCOLUMMNS(
    SUMMARIZECOLUMNS(
        '<tablename x>'
        , <grouping column_1>
        , <grouping column_n>
    )
        , "<new column name_1>" , <expression_1>
        , "<new column name_n>" , <expression_n>
)
```

This pattern is described already in great detail by the awesome people of sqlbi.com (link to the article: www.sqlbi.com/articles/best-practices-using-summarize-and-addcolumns/).

The following mentions some noteworthy peculiarities that should not be forgotten. Listing 7-10 shows the DAX statement to create the table.

Listing 7-10. Table with Summarize

```
Table with Summarize =
ADDCOLUMNS(
    SUMMARIZE(
        'Fact Sale'
        , 'Dimension City'[Sales Territory]
        , "sum inside" , SUM('Fact Sale'[Quantity])
    )
    , "sum outside" , SUM('Fact Sale'[Quantity])
    , "using a measure" , [Total quantity]
)
```

Figure 7-7 shows the table created by the preceding DAX statement.

Sales Territory	sum inside	sum outside	using a measure
External	114537	8950628	114537
Far West	1028670	8950628	1028670
Great Lakes	1047816	8950628	1047816
Mideast	1336832	8950628	1336832
New England	415657	8950628	415657
Plains	1203897	8950628	1203897
Rocky Mountain	582287	8950628	582287
Southeast	1969267	8950628	1969267
Southwest	1251665	8950628	1251665
Total	**8950628**	**80555652**	**8950628**

Figure 7-7. *Table with SUMMARIZE*

To understand what's going on exactly, it is necessary to remember that both functions SUMMARIZE and ADDCOLUMNS are considered to be so-called table iterator functions, but in contrast to SUMMARIZE, ADDCOLUMNS does not perform a context transition and does not create a filter context from the current row of the iteration. This explains why all the row values inside the column `sum outside` are matching the Total row value. As there is no filter context, all the rows from the table iterator are used.

Of course, this effect does not occur if a measure is used to calculate the column values of column `using a measure`. A measure always performs a context transition – transforming a row context into a filter context.

SUMMARIZE vs. GROUPBY

GROUPBY has been introduced to DAX much later, and for this it is not that often used, at least not throughout the community, meaning `htttps://community.powerbi.com`. But this disregard is not justified; this is for two reasons:

- Performance – In certain scenarios, GROUPBY will perform faster than SUMMARIZE. This comes especially true if the number of grouped rows is small (as always, what can be considered a small dataset depends on a lot of factors).

- Advanced Groupings – GROUPBY allows to group "Calculated Columns." This can become very important in complex measures.

Note   This section is referencing the report pages "Nested Ranking," "Ranking within groups SUMMARIZE," and "Ranking within groups GROUPBY."

To compare the performance between both functions, three measures have been defined. For the base measure _r, see Listing 7-11.

Listing 7-11. _r

```
_r =
RANKX(
    CALCULATETABLE(
        SUMMARIZE(
            'Fact Sale'
            , 'Dimension City'[Sales Territory]
            , 'Dimension City'[State Province]
        )
        , ALL('Dimension City'[State Province])
    )
    , [Total quantity]
    ,
    , DESC
)
```

As RANKX by itself is not the easiest one and also not the fastest table iterator function, RANKX is used to put some stress to the underlying data model. What the measure does is this: Within each Sales Territory, the related values of the column State Province are ranked by the measure Total Quantity. Figure 7-8 demonstrates this measure.

Sales Territory	Total quantity ▼	_r matrix
Southeast	**1138684**	
Florida	178472	1
Alabama	177333	2
South Carolina	129509	3
Louisiana	106115	4
West Virginia	104443	5
Georgia	95049	6
North Carolina	93585	7
Virginia	72417	8
Kentucky	68313	9
Tennessee	41967	10
Mississippi	40370	11
Arkansas	31111	12
Mideast	**901870**	
Pennsylvania	359584	1
New York	298604	2
New Jersey	158138	3
Maryland	85544	4
Delaware		5
District of Columbia		5
Total	**5667611**	

Figure 7-8. *The nested ranking*

As this chapter is about using measures to filter a result, two additional measures
have been defined:

- SUMX GroupBy Filtered

- SUMX Summarize Filtered

As both measures are almost identical, only the listing for the first one is displayed
here. See Listing 7-12.

Listing 7-12. SUMX GroupBy Filtered

```
SUMX GroupBy Filtered =
SUMX(
    FILTER(
        ADDCOLUMNS(
            GROUPBY(
                'Fact Sale'
                , 'Dimension City'[Sales Territory]
                , 'Dimension City'[State Province]
            )
            , "q" , [Total quantity]
            , "r" , [_r]
        )
        , [r] <= 2
    )
    , [q]
)
```

What this measure does is straightforward. A summary table is composed using GROUPBY, and Calculated Columns are added to the table using the ranking measure _r. Finally, just the top two states are kept for each sales territory.

In a slight variance to the measure SUMX GroupBy Filtered, the second measure SUMX Summarize Filtered uses the function SUMMARIZE to create the summary table.

Figures 7-9 and 7-10 show the fifth iteration for the performance testing. Each iteration consists of the following steps:

- Open the pbix file with the report page "Blank page" activated.

- Start DAX Studio, connect to the pbix file, and activate the "All Queries trace."

- Switch to the Power BI report page of interest, either "Ranking within groups SUMMARIZE" or "Ranking within groups GROUPBY."

- Select the available slicer values one after the other.

- Repeat the preceding steps for the other report page.

StartTime	Type	Duration	User	Database	Query
04:25:39	DAX	50	tmart	CH 7 - Fil...	DEFINE VAR __DS0FilterTable =
04:25:37	DAX	51	tmart	CH 7 - Fil...	DEFINE VAR __DS0FilterTable =
04:25:34	DAX	52	tmart	CH 7 - Fil...	DEFINE VAR __DS0FilterTable =
04:25:32	DAX	58	tmart	CH 7 - Fil...	DEFINE VAR __DS0FilterTable =
04:25:23	DAX	46	tmart	CH 7 - Fil...	DEFINE VAR __DS0FilterTable =
04:25:23	DAX	8	tmart	CH 7 - Fil...	DEFINE VAR __DS0FilterTable =

Figure 7-9. *SUMMARIZE – filtered nested ranking*

StartTime	Type	Duration	User	Database	Query
04:55:38	DAX	50	tmart	CH 7 - Fil...	DEFINE VAR __DS0FilterTable =	
04:55:35	DAX	42	tmart	CH 7 - Fil...	DEFINE VAR __DS0FilterTable = '
04:55:31	DAX	40	tmart	CH 7 - Fil...	DEFINE VAR __DS0FilterTable = '
04:55:28	DAX	63	tmart	CH 7 - Fil...	DEFINE VAR __DS0FilterTable = '
04:55:21	DAX	43	tmart	CH 7 - Fil...	DEFINE VAR __DS0FilterTable = '
04:55:21	DAX	6	tmart	CH 7 - Fil...	DEFINE VAR __DS0FilterTable =	

Figure 7-10. *GROUPBY – filtered nested ranking*

Both figures show that there is a little advantage of the GROUPBY-based filtering, but it's mandatory to test both functions SUMMARIZE and GROUPBY with the dataset in question.

Binning and the power of GROUPBY

Much has been written about dynamic binning, and more has been said, but more will be said in the future. This means this will not be the last writing about dynamic binning. But as GROUPBY is used, at least there is new twist.

Binning does mean things will be sorted into predefined buckets. Normally these buckets have a lower bound and an upper bound. This example is no exception. Counting the things inside the buckets can reveal tremendous insights, if the size of the buckets has been wisely chosen. Please be warned, the bucket size for this example has been chosen without wisdom. It is also important to notice that the aggregation function is not limited to COUNT.

Note   This section is referencing the report page "Binning."

Figure 7-11 shows the content of the table bins that have been created based on the value distribution of the column Sale Key inside the table Fact Sale.

bins	▼	min	▼	max	▼	binid	▼
0 - 5			0		5		1
6 - 15			6		15		2
16 - 25			16		25		3
26 - 50			26		50		4

Figure 7-11. *The bins*

Figure 7-12 shows the combination of the table bins, the measure, and a ribbon chart.

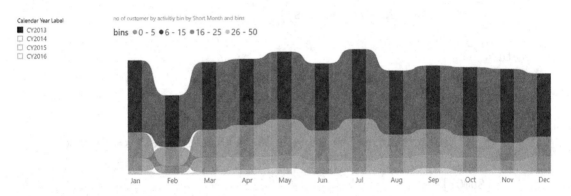

Figure 7-12. *Bins visualized*

Listing 7-13 shows the measure binning using group by.

Listing 7-13. binning using GROUPBY

```
binning using group by =
var abt =
    GROUPBY(
        ADDCOLUMNS(
            VALUES('Fact Sale'[Customer Key])
            ,"NoOfActivities",[No Of Activities]
        )
```

```
        ,[NoOfActivities]
        ,"SumOf",SUMX(CURRENTGROUP(),[NoOfActivities])

    )
return
SUMX(
    VALUES('bins')
    ,
    var binMin = 'bins'[min]
    var binMax =
        IF(
            NOT(ISBLANK('bins'[max]))
            ,'bins'[max]
            ,MAXX(abt,[NoOfActivities])
        )
    return
    SUMX(
        FILTER(
            abt
            ,[NoOfActivities] >= binMin && [NoOfActivities] <= binMax
        )
        ,[SumOf]
    )
)
```

The preceding measure consists of two parts: The first creates the grouping and stores this grouping to the table variable abt. The second part maps the table to the table bins. As this table is not related to the data model, this mapping has to happen dynamically.

The first part, the GROUPBY, is also using the measure No Of Activities. To better understand what is happening, a simple chart is used to explain the intricate workings of the first part of the preceding measure.

Figure 7-13 visualizes the concept of GROUPBY using CURRENTGROUP.

Figure 7-13. *GROUPBY, a conceptual visualization*

The "Innertable" is composed of just two columns, one being the column Customer Key (be aware that the column is used from the table Fact Sale and not from the table Dimension Customer). ADDCOLUMNS is used to add the measure No Of Activities. This measure just counts the number of activities. This table is created by very familiar functions, VALUES and ADDCOLUMNS.

Maybe some of the readers will feel a magical moment when it starts to sink in that the GROUPBY part summarizes a Calculated Column. In combination with the function CURRENTGROUP (this function can be compared to the PARTITON BY clause of the OVER statement from T-SQL), almost a new table has been created with more than one column. Unfortunately, this table cannot be used currently inside visuals, without further precautions, hence the mapping to the unrelated table that represents the buckets.

During the second part, the abt table is mapped to the bins table. This is done using a nested table iteration. The first iteration loops across all the bins, and the second iteration maps all rows of the grouped abt table in accordance to its lower and upper bounds to the bins.

Summary

Using measures to filter datasets, and manipulating the filter expressions of CALCULATE, is not an easy path, especially as there is currently no concise writing that DAX functions can be used exactly. This book is also no exception. DAX is still in a flux, but using measures to filter is rewarding. That is for sure.

PART III

DAX to Solve Advanced Everyday Problems

Using DAX to Solve Advanced Reporting Requirements

Introduction

Until now, DAX has been used to calculate measures or Calculated Columns, aiming to create and extract insights from the data model. Even if this is the most obvious part of using DAX, it's not the only place where DAX can be used. Power BI is a data-driven reporting tool that enables its users to explore large amounts of data; Power BI is also a tool that allows to share reports and collaborate with colleagues. This makes it necessary to create reports and dashboards that are easy to use by its consumers. It's almost mandatory that the report users or the report consumers that have been built by other users are as easy and quick to access as possible. Here, access is not used from a technical point of view. Instead, access is meant from a usability point of view. As soon as the business analyst is going to share his findings, he has to make sure that the recipient of the report can fully concentrate on his main task, extracting insights from the underlying data. Besides creating measures and Calculated Columns, DAX can also be used to create more dynamic reports.

© Philip Seamark, Thomas Martens 2019
P. Seamark and T. Martens, *Pro DAX with Power BI*, https://doi.org/10.1007/978-1-4842-4897-3_8

Some simple but not less powerful DAX

Note   Throughout this section, the Power BI file "CH08 – unrelated tables.pbix" is used.

With the April 2019 release of Power BI Desktop, the feature "Conditional features for visual titles" has been introduced. At first glance, this feature seems to be a minor one, but this feature adds tremendously to the readability of reports and also adds its part in conveying insight.

As for measures that are just changing formatting properties of certain visuals, it is a good idea to separate them from the usual measures. This separation can be achieved by creating a dedicated measure table. As measures are independent of any table at least from a technical point of view, it seems to be a good idea to assign "formatting" measures to a dedicated table.

Creating a measure table

A measure table can be created by simply adding a table using a manual table using Enter Data from the Home menu (see Figure 8-1).

Figure 8-1.   *Create a measure table – Enter Data*

As the content will be deleted after the first measure has been assigned, the content is not important as the next screenshot shows (see Figure 8-2). The name of the table should be chosen carefully as it will provide information about the content.

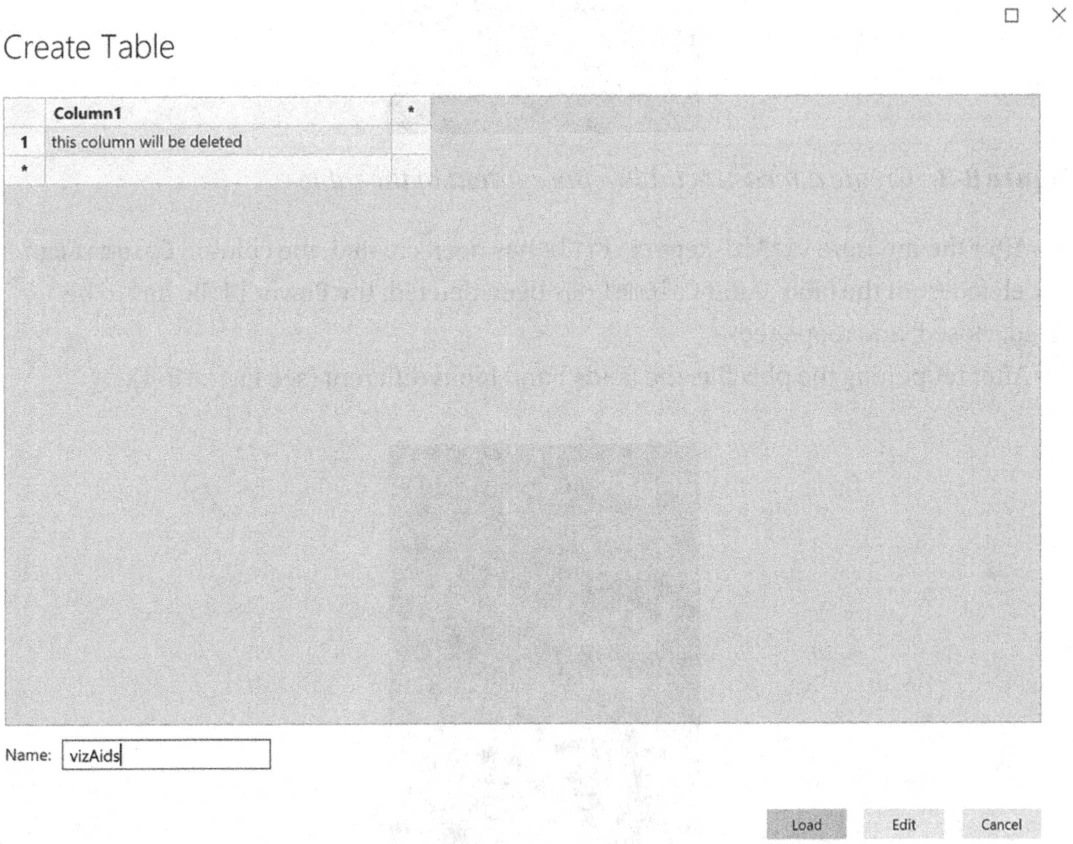

☐   ✕

Create Table

	Column1	✱
1	this column will be deleted	
✱		

Name: vizAids

Load   Edit   Cancel

Figure 8-2. *Create a measure table – the content*

To complete the creation of the measure table, a very simple measure is created using the DAX statement from Listing 8-1.

Listing 8-1. A very simple measure

```
vizAid chart title = "This is the title of a chart"
```

Figure 8-3 shows the content of the newly created table vizAids.

Figure 8-3. *Create a measure table – the content of the table*

After the measure `vizAid Report title` has been created, the column `Column1` can be deleted from the table. After `Column1` has been deleted, the Power BI file has to be saved, closed, and reopened.

After reopening the pbix file, the fields band looks different (see Figure 8-4).

Figure 8-4. *Create a measure table – the measure table*

It has to be noticed that now the table is the first table in the fields listing and that also the symbol now is that of the measure.

Note   Don't assign all of your measures to a dedicated measure table. The Q&A relies on the table assignment of measures as it simplifies discovering relationships to other tables. Assigning measures to a dedicated measure table may impact the Q&A experience.

Using a measure to create a dynamic visual title

This chapter explains how the measure can be used as the title of a chart that is used on the report page "Dynamic axis content."

Enable the title for the chart (see Figure 8-5).

Figure 8-5.   *Using a formatting measure – title*

Next to the property "Title text," an ellipsis button will appear if the mouse comes close. Hitting the ellipsis button allows entering the conditional formatting dialog (see Figure 8-6).

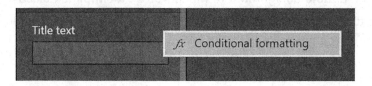

Figure 8-6. *Using a formatting measure – conditional formatting*

A fact to notice is the following: Basically, all properties of a visual can be changed using DAX. As this is just about changing the title dynamically, only one option can be selected. Later on, we will see that more options are available depending on the object that will be formatted. Currently, just the text of the title can be changed using a measure. In a later stage, most likely other properties of the text can also be changed using DAX statements. Figure 8-7 shows how the measure can be assigned, by choosing the option "Format by Field value."

×

Title text

Format by | Field value ▼ | Learn more

Based on field

| vizAid Chart title   ▼ |

| 🔍 Search |

⊿ 📄 vizAids

🔲 vizAid Chart title   Name   'vizAids'[vizAid Chart title]

▸ ▦ Dimension City

▸ ▦ Dimension Customer

▸ ▦ Dimension Date

▸ ▦ Dimension Employee

▸ ▦ Dimension Stock Item

▸ ▦ Fact Sale

▸ ▦ No of Sales Territories

OK   Cancel

Figure 8-7. *Using a formatting measure – title text*

Selecting the measure will create the title for the chart (see Figure 8-8).

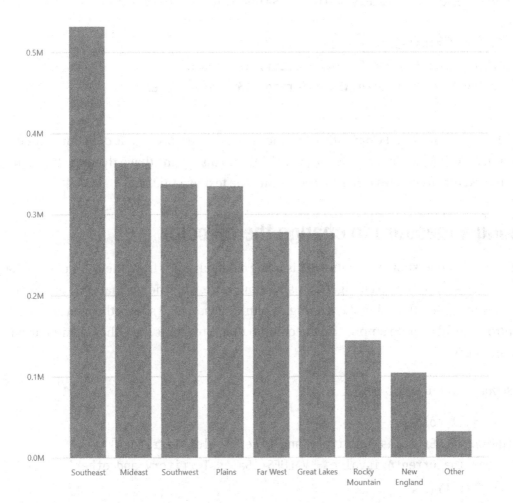

Figure 8-8. *Using a formatting measure – the measure-based title*

As the title is currently not dynamic, another measure will be used, namely, the measure vizAid Chart title dynamic. This measure is shown in Listing 8-2.

Listing 8-2. Slicer title dynamic

```
vizAid Chart title dynamic =
var NoOfTopNSalesTeritory = [No of Sales Territories Value]
return
IF(NoOfTopNSalesTeritory = 1
    ,"Just the Top One Sales Territory is chosen"
    ,"The Top " & NoOfTopNSalesTeritory & " are chosen"
)
```

The measure vizAid Chart title dynamic creates a measure that is dependent on the parameter value coming from the slicer selection. This means that a dynamic title can be composed that integrates other measures into the text of a title.

Using a measure to change the fill color

The column chart shown in Figure 8-8 is using a default color from the color palette that is used. This color can be changed by using a measure. This means that a measure can be used to control the color of each single column. Listing 8-3 shows the measure that controls the fill color to emphasize the column that represents the Other element used on the x-axis.

Listing 8-3. Fill color

```
vizAid fill color =
IF(HASONEVALUE('Sales Territory and other'[Sales Territory])
    ,var theCurrentAxisValue = VALUES('Sales Territory and other'[Sales
    Territory])
    return
    IF(theCurrentAxisValue <> "other", "darkgrey", "darkred")
    ,"darkgrey"
)
```

The measure works like this:

- Check if there is just one value from column 'Sales Territory and other'[Sales Territory] that is selected, using the DAX function HASONEVALUE.

- If HASONEVALUE returns FALSE, this means a single value of the column is not present in the current filter context. The string "darkgrey" is returned.

- IF HASONEVALUE returns TRUE, this means there is a single value present in the current filter context. A second IF function checks if the single value equals the string other or not. If the value equals other, then the string "darkred" is returned; otherwise, the string "darkgrey" is returned.

Figure 8-9 shows the chart after the measure vizAid fill color has been applied to determine the fill color.

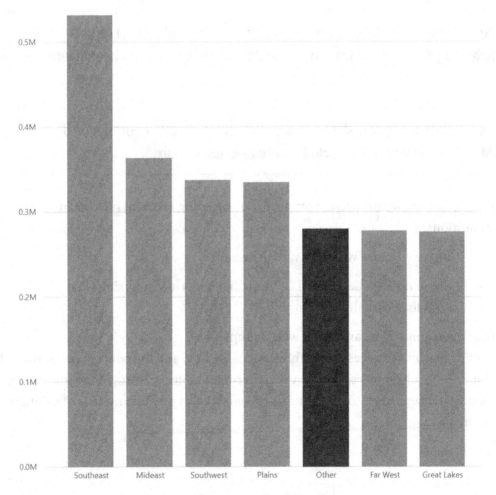

Figure 8-9. *Using a formatting measure – the fill color*

Unrelated tables

Until now, each chapter of this book dealt with a certain aspect of using DAX – table iterators, filtering, and so on. But now there is a more complex task to solve.

Task   Show the Top N Sales Territory (by Quantity) as a column chart, and besides that, all that remain (those who are not counting for the Top) have to be grouped as "Other."

To solve this task, it's necessary to always remember the following fundamental rule.

Rule   It's not possible to visualize a value on an axis of a chart or as a row or column header that's not present in the data model or that has been filtered out.

Note   Unless otherwise stated, this section refers to the report page "Using SUMX" of the Power BI file "CH08 – unrelated tables.pbix."

For this reason, we have to decompose the preceding task and have to answer two basic questions:

- Can we visualize what we want to visualize?

- Can we create a Calculated Column that helps in visualizing what needs to be visualized?

Both questions can be answered with a simple No.

For this reason, it's necessary to think out of the box, and the concept of "unrelated table" may seem obscure, especially remembering Chapter 2, "Data Modeling." But as soon as this concept starts to be natural, unrelated tables provide myriads of solutions to even more tasks.

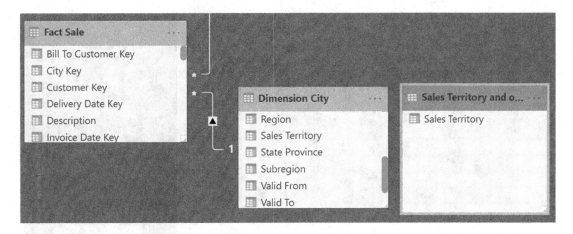

Figure 8-10. *The unrelated table*

Figure 8-10 shows the unrelated table `Sales Territory and other - unrelated`. Using this table on the axis of a chart or as a row or column header allows to visualize elements that are not present in the data model or that have been filtered out by selecting values inside a slicer.

Listing 8-4 shows how the table is created using DAX.

Listing 8-4. The unrelated table

```
Sales Territory and other - unrelated =
UNION(
    DISTINCT('Dimension City'[Sales Territory])
    ,ROW("Sales Territory","Other")
)
```

This table now can be used to visualize a member called `Other`. Unfortunately, solving the first question "Can we visualize what we want to visualize?" comes with a price. As this table is not related to any other table of the data model, it's necessary to map the matching values. This match has to accomplish the task of mapping the values from the column Sales Territory of the table Dimension City to the corresponding member in the unrelated table. Using the Quantity column from the table Fact Sale will not create what we are looking for. Figure 8-11 shows what happens if the unrelated table is used in combination with the `Quantity` column.

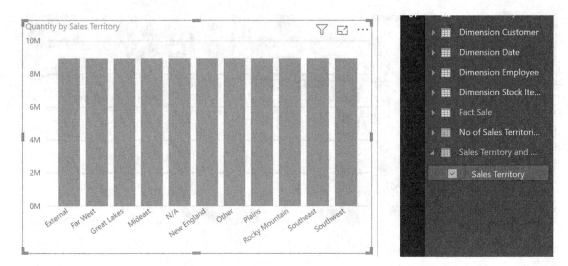

Figure 8-11. *The unrelated table – the unwanted result*

As the column used on the x-axis is coming from an unrelated table, all the values are the same.

To overcome this problem, it's possible to use one of the iterator functions like SUMX. Chapter 6, "Iterators," already explained and demonstrated the workings of an iterator function – creating a measure that iterates over the table that contains the values from the Sales Territory column of the unrelated table. Listing 8-5 defines the measure that maps the Quantity value to the corresponding Sales Territory of the unrelated table.

Listing 8-5. Simple mapping to an unrelated table

```
_simple mapping to the unrelated table =
SUMX(
    VALUES('Sales Territory and other - unrelated'[Sales Territory])
    ,
    var SalesTerritoryFromUnrelated = 'Sales Territory and other -
unrelated'[Sales Territory]
    return
    CALCULATE(SUM('Fact Sale'[Quantity]), 'Dimension City'[Sales Territory]
= SalesTerritoryFromUnrelated)
)
```

Figure 8-12 shows the chart as in the preceding text, but now also the measure from Listing 8-6 is used.

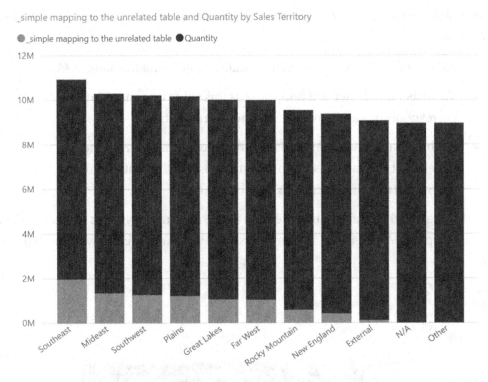

Figure 8-12. *The unrelated table – simple mapping*

Figure 8-12 shows that the corresponding values are mapped. The mapping works like this:

- Iterate across each row of the table parameter. For this, it's necessary to use the VALUES function as this function returns a table of the referenced column within the current filter context. As the axis value by itself creates a filter context, for this the table just contains one row.

- Calculate the Quantity by using the unrelated Sales Territory value as filter of the Dimension City table.

Caution   It's essential to notice that due to the usage of SUMX in combination with VALUES, the calculated Quantity is just mapped, and there are no intrinsic summations of multiple Sales Territories. This is because of the fact that each table (returned by VALUES) just contains a single row.

Figure 8-12 unfortunately reveals three additional aspects that have to be considered for a final solution:

- Axis value will show up, even if no value exists in the fact table N/A.

- The axis value Other will disappear as this value does not find a corresponding value inside the Dimension City table.

- As the iteration happens across all the column values from the unrelated table, a final solution has to consider that not all Sales Territories exist in the current filter context.

For this reason, the final solution, defined in the measure _dynamic content SUMX, is a more complex DAX measure. Figure 8-13 shows the column chart with the expected result.

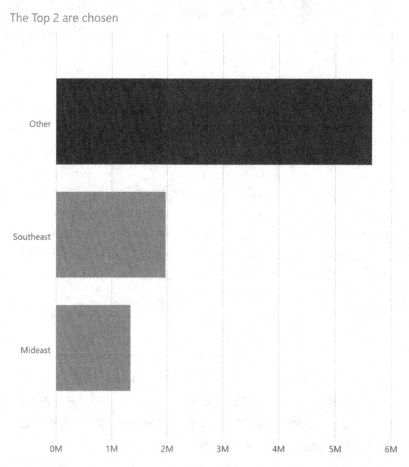

Figure 8-13. *The expected result*

The definition of this measure is shown in Listing 8-7. The rest of this chapter explains the sections in more detail.

As the measure relies on a measure that uses the iterator function RANX, this measure will be shown in Listing 8-6.

Listing 8-6. The final solution – the ranking

```
_Rank =
RANKX(
    ALLSELECTED('Dimension City'[Sales Territory])
    , CALCULATE([Total Quantity])
    ,
    , DESC
)
```

Listing 8-7. The final solution – using SUMX

```
_dynamic content SUMX =
var NoOfParameter = [No of Sales Territories Value]
var SelectedSalesTerritories = VALUES('Dimension City'[Sales Territory])
var tableWithRank =
    FILTER(
        ADDCOLUMNS(
            SelectedSalesTerritories
            ,"rank",[_Rank]
        )
        ,[rank] <= NoOfParameter
    )
var topNTable =
    SELECTCOLUMNS(
        tableWithRank
        ,"axisvalue",[Sales Territory]
    )
var tableOther = EXCEPT(DISTINCT(SelectedSalesTerritories),topNTable)
var measureValueOther = CALCULATE([Total Quantity],tableOther)
```

```
var tableToIterate =
    ADDCOLUMNS(
        UNION(
            INTERSECT(DISTINCT('Sales Territory and other -
            unrelated'),topNTable)
            ,ROW("axisvalue","Other")
        )
        ,"theValue"
        ,var thisCategory = [Sales Territory]
        return
        IF(thisCategory <> "Other"
            , CALCULATE([Total Quantity], 'Dimension City'[Sales Territory]
            = thisCategory)
            , measureValueOther
        )
    )

return
SUMX(
    FILTER(
        tableToIterate
        ,'Sales Territory and other - unrelated'[Sales Territory] in
        VALUES('Sales Territory and other - unrelated'[Sales Territory])
    )
    ,[theValue]
)
```

The variable definition

```
var NoOfParameter = [No of Sales Territories Value]
```

is a very simple line. This line stores the value of the measure [No of Sales
Territories Value]. This value represents the value of the parameter slicer and is
used to filter a table that only contains the Sales Territories that correspond to the Top N
selection by the user of the report page.

The variable definition

```
var SelectedSalesTerritories = VALUES('Dimension City'[Sales Territory])
```

is also very simple, but very very important. This line considers already filtered Sales Territories. Using VALUES with the column reference to the column Sales Territory from the table that is related to the fact table makes sure that only Sales Territories within the current filter context are considered. This will become obvious if you try different selections from the available slicers "State Province" or "Sales Territory." Both of these slicers are using columns from the related table. Figure 8-14 shows that filtered Sales Territories are considered in the current filter context.

Note   You have to be aware that the State Province slicer is only used to "reduce" the available Sales Territories. The impact of values is not considered.

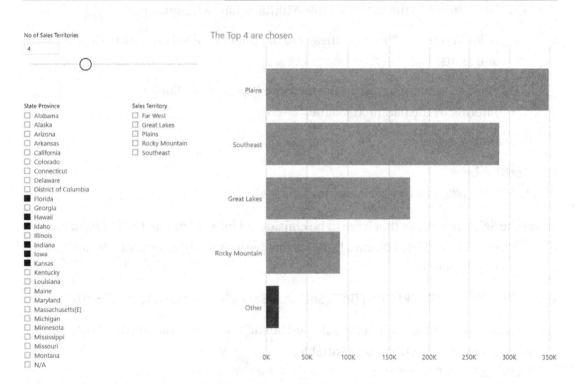

Figure 8-14. *The unrelated table – filtered Sales Territories*

The variable definition

```
var tableWithRank =
    FILTER(
        ADDCOLUMNS(
            SelectedSalesTerritories
            ,"rank",[_Rank]
        )
        ,[rank] <= NoOfParameter
    )
```

stores the Sales Territories that have to be considered as TOP N into a dedicated table variable with a rank lesser than or equal to the selected parameter value. As ADDCOLUMNS is one of the table iterator functions, it can be used to assign a rank to it by using the measure _Rank. The resulting table is filtered by the parameter value. At this moment, the table that is stored to the variable tableWithRank has two columns:

- Sales Territory – The data lineage to the original table Dimension City still exists.

- rank – A column based on the measure [_rank] that allows the filtering by the parameter value.

The variable definition

```
var topNTable =
    SELECTCOLUMNS(tableWithRank,"axisvalue",[Sales Territory])
```

stores the Sales Territories that have to be considered into a separate table. This table now contains only a single column, but data lineage to the original column is not broken.

The variable definition

```
var tableOther = EXCEPT(DISTINCT(SelectedSalesTerritories), topNTable)
```

extracts the Sales Territories that are selected but do not match the threshold of the Top N and stores them into a dedicated variable.

The variable definition

```
var measureValueOther = CALCULATE([Total Quantity],tableOther)
```

stores the value for the Other Sales Territories. The table variable tableOther is used to apply a filter before the Quantity is calculated.

The variable definition

```
var tableToIterate =
    ADDCOLUMNS(
        UNION(
            INTERSECT(DISTINCT('Sales Territory and other - unrelated'),
            topNTable)
            ,ROW("axisvalue","Other")
        )
        ,"theValue"
        ,var thisCategory = [Sales Territory]
        return
        IF(thisCategory <> "Other"
            ,CALCULATE([Total Quantity], 'Dimension City'[Sales Territory]
            = thisCategory)
            ,measureValueOther
        )
    )
```

stores the table that will be used as a table in the SUMX function. Using INTERSECT creates a table that only contains the Sales Territories that are present in the table variable topNTable. This table is combined with a single-row table created by using the DAX function ROW. Both tables topNTable and the single-row table are combined by using the function UNION. This table now references the unrelated table.

The Quantity will be calculated and stored in the column theValue by using the current row value (column Sales Territory) as a filter for the related Dimension City table, but this time the table only contains the Sales Territories that correspond to the available Sales Territories, available in the original filter context. This calculation only happens if the value of the Sales Territory does not equal Other. If the value equals Other, the previously stored value is used (variable measureValueOther).

Finally, the following lines

```
SUMX(
    FILTER(tableToIterate
        ,'Sales Territory and other - unrelated'[Sales Territory] in
        VALUES('Sales Territory and other - unrelated'[Sales Territory]))
    ,[theValue]
)
```

iterate across a table that is returned by a FILTER statement. Basically, SUMX is used to iterate across a table to map a value. This is similar to the mapping demonstrated in Listing 8-5. To avoid that all the rows that are present in the final table are summed for all values that are used on the axis of the column chart, the FILTER function is used, whereas the filter criteria (the implicit filter) coming from the visual is "injected" into the formula by using the VALUES function. The filter function makes sure that the iterated table only contains one single row.

Sorting of the Other member using a tooltip

Depending on the number provided as Top N criteria, the dynamically calculated value that represents the remaining Sales Territories, the Other member appears on the top, somewhere in between or at that bottom. This depends solely on the number of Sales Territories selected by the parameter value. This behavior can be exactly what the users of the report are looking for. But it can also be possible that the axis value representing Other always has to appear at the bottom of the bar chart.

This can be achieved by defining a simple measure, simple in comparison to the last. The measure vizAid sort rank is used to ensure that the axis value Other will always appear at the bottom of the chart. Listing 8-8 the DAX statement of this measure.

Listing 8-8. The final solution – the sorting

```
vizAid sort rank =
var NoOfParameter = [No of Sales Territories Value]
var tableSelectedSalesTerritories = VALUES('Dimension City'[Sales
Territory])
var tableWithRanks =
    FILTER(
        ADDCOLUMNS(
            tableSelectedSalesTerritories
            ,"rank",[_Rank]
        )
        ,[rank] <= NoOfParameter
    )
```

```
var TableToSort =
    UNION(
        SELECTCOLUMNS(
            tableWithRanks
            ,"axis", [Sales Territory]
            ,"rank", [rank]
        )
        ,ROW("axis", "Other", "rank", IF(COUNTROWS(tableSelectedSales
        Territories) > NoOfParameter, NoOfParameter + 1, BLANK())))
    )

return

sumx(
    FILTER(
        TableToSort
        ,[axis] in VALUES('Sales Territory and other - unrelated'[Sales
        Territory])
    )
    ,[rank]
)
```

A lot of the DAX from Listing 8-8 has already been explained. For this reason, only a short summary is given:

- Store the Top N parameter value to a variable.

- Store the selected Sales Territories that are present in the current filter context.

- Store the Top N Sales Territories to a table.

- Create a table with the Other axis row by using UNION(..., ROW(...)).

- To make sure that the axis value Other always will appear at the bottom, the variable NoOfParameter is incremented by one.

- Map the filtered table to the corresponding axis value.

Basically, it's not the definition of the measure that seems difficult but instead the application of the measure to achieve the proper ordering of the bars. For this reason, one has to know that the ordering of the axis values can also be controlled by measures that are used as tooltips. Figure 8-15 shows how the sorting has to be configured.

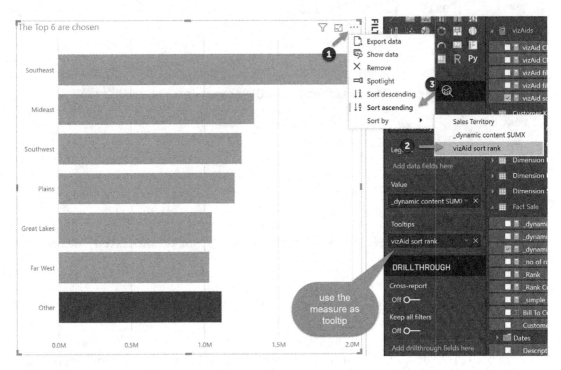

Figure 8-15. *The unrelated table - ordered by a tooltip*

The measure has to be assigned to the tooltips of the bar chart. Then the sorting options of the visual have to be configured accordingly.

Dynamic measure selection using a slicer

Besides the dynamic grouping of remaining column values into an axis value called Other, another very common requirement is the application of a measure using a slicer selection.

Task   Create a report page that provides the possibility of selecting a measure from a dropdown that will be applied to all the visuals on the report page.

The implementation of the preceding task is demonstrated on the report page "USING SUMX – final."

Note   This section refers to the report page "Using SUMX – final" of the Power BI file "CH08 – unrelated tables.pbix".

Similar to the unrelated table containing the Sales Territories used in the previous chapter, this solution starts with the creation of an unrelated table. This time, this table contains only the measures that the user can choose (see Listing 8-9).

Listing 8-9. The unrelated measure table

```
Measure Table =
UNION(
    ROW("Slicer" , "Profit")
    , ROW("Slicer" , "Quantity")
)
```

In addition to the unrelated table, it's also necessary to create a measure. The definition of this measure is shown in Listing 8-10.

Listing 8-10. The selected measure

```
Measure Table SelectedValue =
var theSelectedMeasure = SELECTEDVALUE('Measure Table'[Slicer], "Default")
return
SWITCH(theSelectedMeasure
    , "Profit" , [Total Profit]
    , "Quantity" , [Total Quantity]
    , [Total Quantity]
)
```

Depending on the selected value from the slicer, the measure will evaluate accordingly. If none oder more values are selected from the slicer, the evaluation defaults to the measure Total Quantity.

As the measure _dynamic content SUMX is using the ranking measure _Rank to select the Sales Territories accordingly, it's also necessary to adjust the existing measure.

For comparison purposes, a second ranking measure has been defined. Listing 8-11 shows the definition of this measure.

Listing 8-11. The final solution – Rank final

```
_Rank final =
RANKX(
    ALLSELECTED('Dimension City'[Sales Territory])
    , CALCULATE([Measure Table SelectedValue])
    ,
    , DESC
)
```

The measure _Rank final references the measure Measure Table Selected Value. This reference ensures that the measure selected from the slicer will be evaluated in the present filter context.

Note   Actually, the slicer selection does not really select a measure; instead, the measure "reacts" to the slicer selection.

The preceding measures (Listings 8-10 and 8-11) are used in the measure _dynamic content SUMX final shown in Listing 8-12. As this measure is basically a copy of the measure that has been explained in great detail, here only the lines are highlighted where the preceding measure is used.

Listing 8-12. The final solution – the final measure

```
_dynamic content SUMX final =
var NoOfParameter = [No of Sales Territories Value]
var SelectedSalesTerritories = VALUES('Dimension City'[Sales Territory])
var tableWithRank =
    FILTER(
        ADDCOLUMNS(
            SelectedSalesTerritories
            , "rank",[_Rank final]
        )
```

```
        , [rank] <= NoOfParameter
    )
var topNTable =
    SELECTCOLUMNS(
        tableWithRank
        , "axisvalue" , [Sales Territory]
    )

var tableOther = EXCEPT(DISTINCT(SelectedSalesTerritories),topNTable)
var measureValueOther = CALCULATE([Measure Table SelectedValue],tableOther)

var tableToIterate =
    ADDCOLUMNS(
        UNION(
            INTERSECT(DISTINCT('Sales Territory and other - unrelated') ,
            topNTable)
            , ROW("axisvalue" , "Other")
        )
        , "theValue"
        , var thisCategory = [Sales Territory]
        return
        IF(thisCategory <> "Other"
            , CALCULATE([Measure Table SelectedValue] , 'Dimension
            City'[Sales Territory] = thisCategory)
            , measureValueOther
        )
    )

return
SUMX(
    FILTER(
        tableToIterate
        , 'Sales Territory and other - unrelated'[Sales Territory] in
        VALUES('Sales Territory and other - unrelated'[Sales Territory])
    )
    , [theValue]
)
```

Caution   We always have to keep in mind that each feature that helps to make a report more accessible to the user has a price, performance! For this reason, we have to implement features carefully.

Color: Use color to emphasize the meaning of data

Microsoft has announced that DAX can be used to manipulate each property of a visual. The ability to change properties of a visual by using DAX will add tremendous power to report design. This is simple because visual properties can help to emphasize the content of a report. As the implementation will span a couple of Power BI Desktop releases, this is not the place to cover all the possibilities. Some of the visual properties already have been used in previous chapters. For this reason, this chapter solely focuses on the implementation of a heatmap that will allow to use custom colors for certain products. The announcement can be found here: https://docs.microsoft.com/en-us/ business-applications-release-notes/April19/business-intelligence/power-bi-desktop/expression-based-formatting.

Note   Throughout this section, the Power BI file "CH08 – color gradients.pbix" is used.

Task   Create a table heatmap that utilizes a configured color for each product; if no color is configured, use a default color. The color has to be visualized on a gradient scale, where a higher percentage value will result in a darker color, meaning the value 100% will show the configured product color.

The result: A DIY heatmap (a more complex heatmap)

This chapter contains some words about color theory, converting hex values into RGB values and back again, and also some DAX. This chapter will start with the result to reveal what's possible using the combination of Power BI and DAX. Figure 8-15 will reveal a first glance to the heatmap.

some rows	0	1	2	3	4	5	6	7	8	9
0	0.01	0.02	0.03	0.04	0.05	0.06	0.07	0.08	0.09	0.10
1	0.11	0.12	0.13	0.14	0.15	0.16	0.17	0.18	0.19	0.20
2	0.21	0.22	0.23	0.24	0.25	0.26	0.27	0.28	0.29	0.30
3	0.31	0.32	0.33	0.34	0.35	0.36	0.37	0.38	0.39	0.40
4	0.41	0.42	0.43	0.44	0.45	0.46	0.47	0.48	0.49	0.50
5	0.51	0.52	0.53	0.54	0.55	0.56	0.57	0.58	0.59	0.60
6	0.61	0.62	0.63	0.64	0.65	0.66	0.67	0.68	0.69	0.70
7	0.71	0.72	0.73	0.74	0.75	0.76	0.77	0.78	0.79	0.80
8	0.81	0.82	0.83	0.84	0.85	0.86	0.87	0.88	0.89	0.90
9	0.91	0.92	0.93	0.94	0.95	0.96	0.97	0.98	0.99	1.00

Figure 8-16. Color gradient – first glance

Probably (very very likely) Figure 8-16 just shows a chart with different shades of gray. Just open the pbix file "CH08 – color gradient.pbix", select the report page "Matrix visual," and select one of the products from the slicer.

Note   The used data model may not meet your expectations regarding a data model. Consider this pbix as a sketchbook for the development of a heatmap. Each table will be explained in detail through the following chapters.

Note: All the following chapters will use lines from the measure factored color background.

Some words about color theory

As this is a book about DAX and not about color theory, only the least viable information is used.

Hexcolor and RGB

Quite often, a hex value is used to encode color information, for example, the hexcode #b34529 encodes a shade of red, brown, etc. To start playing with colors, there are many online resources available, for example, www.w3schools.com/colors/colors_picker.asp.

Please be aware that the mentioned web site will not help to create awesome reports if used without any experience in information design. Nevertheless, it's good to know that each two characters of the preceding hexcode represents a color from the RGB color wheel:

B3 → R(ed)   → R: 179

45 → G(reen) → G: 69

29 → B(lue)   → B: 49

The following lines of DAX statements show how a given hexcode is transformed into the RGB components:

```
var startcolor = --"#f45f42"
    IF(HASONEVALUE('Dimension Product'[Dim Product])
        ,FIRSTNONBLANK('Dimension Product'[Report Color],0)
        ,"#808080"
    )
var thepercentagevalue = CALCULATE([simple measure])
var startR = mid(startcolor, 2,2)
var startG = mid(startcolor, 4,2)
var startB = mid(startcolor, 6,2)
var resultR =
    var poweroneStartR = LOOKUPVALUE('hex2dec'[dec],hex2dec[hex],
    left(startR,1))
    var powerzeroStartR = LOOKUPVALUE('hex2dec'[dec],hex2dec[hex],
    right(startR,1))
    var intStartR = poweroneStartR * POWER(16,1) + powerzeroStartR *
    power(16, 0)
    return intStartR
```

This is what happens in the preceding lines:

1. Get the configured color for a given product. If no color is configured or no product is filtered, the default color #808080 is used.

2. The R(ed) part is stored into the variable startR ...

3. The integer value is calculated and stored into the variable resultR.

To calculate the integer value, the table hex2code is used to look up the integer values that are decoded by the hex values 0–9 and A–F.

Step 2 is repeated for G and B parts.

Step 3 is repeated for the G and B parts even if just the R(ed) part is visible in the preceding DAX lines.

Figure 8-17 shows the content of the table Dimension Product.

Dim Product	Report Color	just a number
Product A	#7d4f73	1
Product B	#228b22	1
Product C	#003366	1

Figure 8-17. *Color gradient – table Dimension Product*

The content of the table Dimension Product has been created by using the feature "Enter Data." This allows to easily test the heatmap by using default colors from your CI. Just open Power Query, select the table Dimension Product, hit the gear icon, and change the colors. Please use the # as the first character and also make sure that you enter a valid color (#zzzzzzz is no valid color). Figure 8-18 shows how the configured product colors can be adjusted.

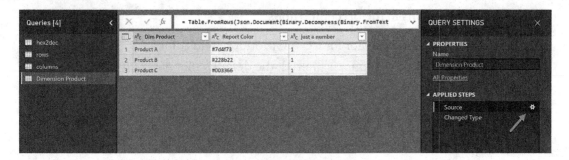

Figure 8-18. *Color gradient - changing the product colors*

Color factorization

Color factorization is necessary to create a gradient of the base color, the color retrieved from the `Dimension Product` table. The following lines show how this happens for the R(ed) part:

```
var factoredColorR =
    var factorizedR = ROUND((resultR - 255) * thepercentagevalue + 255,0)
    var intR = factorizedR
    var divR = intR/16
    var hexPowerR1 = LOOKUPVALUE('hex2dec'[hex],hex2dec[dec],TRUNC(divR))
    var hexPowerR0 = LOOKUPVALUE('hex2dec'[hex],hex2dec[dec],round((divR-
    TRUNC(divR)),0))
    return CONCATENATE(hexPowerR1,hexPowerR0)
```

The preceding algorithm is derived from the Wikipedia article `www.rapidtables.com/convert/color/rgb-to-hsl.html` (providing a theoretical background) and the article `www.rapidtables.com/convert/color/rgb-to-hsl.html` (a more practical approach).

Finally, the three variables are stitched together using this simple line:

```
"#" & factoredColorR & factoredColorG & factoredColorB
```

The measure `thepercentagevalue` is passed to this measure and used to "factorize" the base color in accordance to the value of `thepercentagevalue`.

A final note

What this chapter demonstrates is the following: DAX can be used to develop even complex algorithms. The factorization is a much more complex algorithm than the computation of `Quantity * Price`.

And also the tables, columns, and rows will play a greater part in the development of complex algorithms and in the scaling/testing of DAX statements. A more elaborate usage of these tables can be seen in Chapter 14, "Scale Your Models."

Time Intelligence

Introduction

Back in 1582 when Pope Gregory XIII introduced his Gregorian calendar as an update to the previously used Julian calendar, he didn't have business intelligence reporting in mind. The quirky calendar still included

> Months with an uneven number of days – 28, 29, 30, and 31

> A weekday pattern of 7 days that did not align with calendar years or months

> Leap years

The main intent of the new calendar was to stop Easter moving away from the spring equinox each year, rather than to provide a nice, consistent method of grouping by aligning dates into easy-to-compare blocks.

The new system included a change to the way leap years got calculated. It changed from one that added a day in February every four years to a new version that only included an extra day if the year was divisible by four, except if the year was also divisible by 100 – unless the year was also divisible by 400, in which case a day got added regardless. Makes sense and easy to remember? I'm sure this system got devised by someone with a job title of *project manager*.

The calendar was not universally popular. In Britain in 1752, streets were filled by rioters rejecting the new system. Some historians have suggested the rioters were Protestants who saw the calendar as a Catholic plot, but I suspect it was more to do with the havoc they saw this would have on future generations of data modelers.

Of course, a far more sensible solution would have been to introduce a metric system based around a 10-day week, with a year being 100 days (or 10 weeks). Life expectancy would have more than trebled overnight, which is more than can be said for any medical breakthrough over the ages from discoveries such as vaccines or improved sanitation.

© Philip Seamark, Thomas Martens 2019
P. Seamark and T. Martens, *Pro DAX with Power BI*, https://doi.org/10.1007/978-1-4842-4897-3_9

Moreover, once humanity stops bickering long enough to focus its attention on mastering proper space travel and exploration, the *Seamark calendar* (metric) is inevitable. Confusing constructs such as months and quarters do not have a place in the vastly improved calendar.

Until then, we are stuck with the Gregorian calendar with all its quirks. To make life easier for data modeling and working with DAX, the language provides a set of functions to help organize, categorize, and group days in ways that help make some sense of the modern calendar.

A helpful thing to know about dates in DAX is they get stored as integers inside the DAX engine. To demonstrate this, consider the following calculated table in Listing 9-1 with results shown in Figure 9-1.

The calculation in Listing 9-1 generates a list of numbers between negative five and positive five, incrementing by 1. The series of numbers gets cloned from the first column to the second column before being formatted to display in a date format.

Listing 9-1. A calculated table is showing the same numbers in integer and date format

```
Date Table =
    ADDCOLUMNS(
        GENERATESERIES(-5,5),
        "My Date Format",FORMAT([Value],"Long Date")
        )
```

Value	My Date Format
-5	Monday, 25 December 1899
-4	Tuesday, 26 December 1899
-3	Wednesday, 27 December 1899
-2	Thursday, 28 December 1899
-1	Friday, 29 December 1899
0	Saturday, 30 December 1899
1	Sunday, 31 December 1899
2	Monday, 1 January 1900
3	Tuesday, 2 January 1900
4	Wednesday, 3 January 1900
5	Thursday, 4 January 1900

Figure 9-1. *Shows the output of code in Listing 9-1*

Note the value of zero in the sixth row is assigned to Saturday, 30 December 1899, and calendar days on either side of this date are just representations of a negative or positive difference from this baseline date.

The underlying value of -693,593 represents 1 January 1 AD. According to the DAX interpretation of the Gregorian calendar, this was a Monday, which is an argument to support the case for Monday to be considered the real start of a calendar week, rather than Sunday. ☺

Note   Dates before 30 December 1899 (the negative sequence numbers) are not officially supported as some DAX date/time functions/operates don't work on older dates.

Calculations in DAX throw an error if you try and manipulate a DateTime value to one that is before 1 January 1 AD but do allow you to add nearly 3 million days beyond the baseline date to take you to 31 December 9999. 31 December 9999 happens to be a Friday, which is ideal for what is likely to be a pretty big party. However, keep in mind the metric-based Seamark calendar has probably kicked in long before this.

Every date between -657,435 and 2,958,465 is assigned useful properties like Year, Month, Weekday name, and more, which provide a variety of ways to group days for time-based reporting.

Time Intelligence

What does all this have to do with Time Intelligence? I think it is useful to remember that dates in DAX get treated as numbers and Date Tables are just a sequence of rows of consecutive days that never allow gaps.

Suppose you have a requirement to develop a calculation that reports the value of a metric for every calendar month in 2020 and, for each month, also reports the difference in value between it and the same metric from the previous calendar month. Oh, and on top of this, shows a year-to-date running total that calculates based on values for the last day of each month.

The first challenge is to establish which integers (dates) belong to the year 2020. Applying logic to the known baseline date of zero (30 December 1899), taking into account leap years, you would hopefully work out the dates needed are between 43,831 (1 January 2020) and 44,196 (31 December 2020). Hopefully, the output of this first calculation is a table with 366 days with additional columns showing the different ways each date can get grouped. At least one column shows to which month each row belongs.

It's possible to write this raw calculation in DAX without time intelligence, but it is likely to involve a lot of code and potentially very error-prone. This complex code gets written before having to deal with the additional requirements of period comparison and running totals.

For the year 2020, care gets taken with February due to it having an extra day this year. When needing to calculate a value for the previous month, when the current month is March of 2020, the measure needs to consider all 29 days associated with February and not 28 days as per other years.

Again, this is possible with lots of complex DAX, but fortunately, the suite of Time Intelligence functions come to the rescue and help reduce this complexity by introducing functions that natively understand the Gregorian calendar and all its quirks.

Date Tables

A crucial element for time intelligence functions to work is the presence of a Date Table in the data model. The vast majority of data models I have worked with require an ability to represent and report measures over time, so they require at least one table in them to operate as a Date Table.

A Date Table provides a platform for Time Intelligence functions to work; however, the functions make the following assumptions about any Date Table used.

A Date Table has

> A column using the DATE or DATETIME datatype
>
> Unique values in the preceding column
>
> Contiguous values with no gaps between the oldest and newest values

It's possible to use a Date Table with DAX time intelligence functions that do not conform to all three of the preceding rules, but any results are unreliable as these assumptions are never validated and errors do not get generated.

Date Tables can be constructed using DAX, the query editor, or sourced externally from Power BI Desktop. Ideally, your organization has a well-managed repository of data entities such as an easy-to-access data lake or data warehouse that contains a prebuilt date dataset.

A Date Table should carry more than a single column and include many additional columns that help categorize individual days into useful groupings such as year, month, weekdays, holidays, and so on.

There are some excellent free examples available on the Internet that show how to build Date Tables in DAX. These often start with a core function such as the CALENDAR or CALENDARAUTO, which allows you to either specify a start/end date for the range or pick up these values automatically based on data in your model.

The output is a table that conforms to the three rules specified earlier in this chapter. Listing 9-2 shows a calculated table that contains a single DateTime column that starts on 1 January 2010 and ends on 31 December 2020. The table contains 4,018 rows and has no gaps or duplicates. A sample output of the code at Listing 9-2 is shown in Figure 9-2.

Additional columns can get added to this table by either extending the calculation using the ADDCOLUMNS function or using the New Column button on the ribbon.

Listing 9-2. A Date Table built in DAX using the CALENDAR function

```
Dates =
    CALENDAR(
        DATE(2010, 1, 1) ,
        DATE(2020,12,31)
    )
```

Figure 9-2. *Shows the first seven rows of the output for the calculated table in Listing 9-2*

Once the data model has a Date Table, the Date Table can get linked to other tables that have data to be measured and aggregated by date. The column used to define a relationship can be DateTime or another datatype. However, DAX Time Intelligence functions need to use a column in the Date Table that has a datatype of DateTime.

Note   Always check to see if data in a column used in a relationship to a Date Table includes a time element such as hours, minutes, or seconds other than midnight. These rows get ignored by your time intelligence calculation, and this may not be immediately apparent.

Multiple relationships can get defined between a Date Table and another table in the model. It's also possible to have more than one Date Table in a data model.

Auto DateTime tables

Power BI Desktop comes with a feature that automatically adds Date Tables to new data models. This feature works by automatically creating a hidden table for every column detected in the data model that uses a Date or DateTime datatype.

If you have a data model with five tables and across those tables there happen to be ten columns using a Date data type, then Power BI Desktop adds ten Date Tables to the model that are all hidden. Power BI Desktop also defines a relationship between the date column in your table in the data model and the automatically created hidden Date Table.

This auto created Date Table can be a handy feature for users who are new to data modeling that don't want to manage Date Tables themselves. More advanced users used to the concept of Time Intelligence tend to disable this feature and add/manage Date Tables themselves.

Each automatically created Date Table has seven columns and a single hierarchy. These tables are visible in external tools such as DAX Studio and SSMS when they get connected to a Power BI Desktop data model.

The seven columns are as follows:

Date	[DateTime]	
Day	[Integer]	{1–31}
Month	[Text]	{January–December}
MonthNo	[Integer]	{1–12}
Quarter	[Text]	{Qtr1–Qtr4}
QuarterNo	[Integer]	{1–4}
Year	[Integer]	

The single hierarchy generated in the automatic Date Table is [Year] \ [Quarter] \ [Month] \ [Day].

When this feature is enabled and a date (or DateTime) column gets added to a visualization field, the hierarchy appears in the field box. Individual levels of the hierarchy can be deleted, such as Quarter, while still retaining the drill up/down functionality.

Many of the Quick measures rely on this feature enabled.

In DAX, a particular third level of the fully qualified notation becomes available. This notation takes the form of *'<your table name>'[Your Date Column].[Date]* where the final reference after the dot can get substituted for any of the seven columns from the auto DateTime table.

Given that this is a book about Pro DAX, you may do what I do with every new model and head straight into the options and settings, finding the Data Load section of the CURRENT FILE and unchecking the "Auto date/time" box. Once this is unchecked, the hidden tables all disappear for that specific pbix file.

Time Intelligence functions: The basic pattern

If you have read Chapter 5, "Filtering in DAX," you'll know it's possible to use DAX table expressions as filters. The same chapter also covers the concept of unblocking (overwriting) explicit filters set by items such as slicer selections, axis, and row and column header values.

I like to think of the time intelligence functions as a set of smart, time-aware tables of dates that get used as filters to help identify the correct rows to get considered by the core aggregation functions (SUM, COUNT, AVERAGE, and so on.)

Nearly all the Time Intelligence functions in DAX accept a parameter that needs to be a column of dates. This parameter should be a reference to a column from a Date Table that uses the Date (or DateTime) datatype.

Time Intelligence functions return a single-column table of dates that get used by the calling function to filter and identify rows to get considered by the calculation. Time Intelligence functions usually get called by the CALCULATE function from a <filter> parameter. The dates contained in the table instruct the CALCULATE function which rows to use.

Consider a classic star schema model as in Figure 9-3 that uses two tables from the Wide World Importers dataset, Dimension Date and Fact Sale. A single relationship gets defined between the two tables using the [Date] column from Dimension Date and the Invoice Date Key column from Fact Sale.

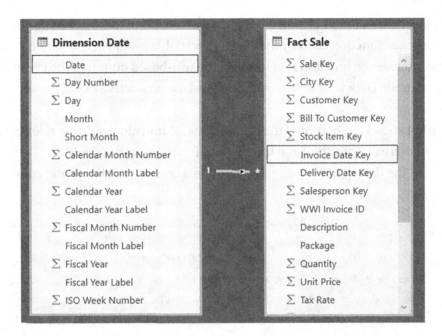

Figure 9-3. *Shows a two-table data model using Wide World Importers data*

Adding the Calculated Column in Listing 9-3 to a Date Table is one way to demonstrate how Time Intelligence functions work. In this example, the column gets added to the Dimension Date Table.

Listing 9-3. A Calculated Column added to a basic Date Table built in DAX using the CALENDAR function

```
TI Function Debug Column =
VAR MyFunct = PREVIOUSMONTH('Dimension Date'[Date])
RETURN
    COMBINEVALUES(
        ", ",
        "Min Date=" & FORMAT(MINX(MyFunct,[Date]),"YYYY-MM-dd"),
        "Max Date=" & FORMAT(MAXX(MyFunct,[Date]),"YYYY-MM-dd"),
        "Rowcount=" & COUNTROWS(MyFunct)
    )
```

When the calculation in Listing 9-3 is added as a column to a Date Table and the Time Intelligence function used by the *MyFunct* variable is adjusted to whichever DAX Time Intelligence function is of interest, the text-based output in the column can be reviewed in the Data view to better understand the characteristics of the selected function.

Figure 9-4 shows the output of the Calculated Column alongside the relevant date for each row. In this example, the Time Intelligence function used is the PREVIOUSMONTH function, and the date column that gets passed to it comes from the [Date] column from the same table.

Date ⛛	TI Function Debug Column ▾
Tuesday, 26 March 2013	Min Date=2013-02-01, Max Date=2013-02-28, Rowcount=28
Wednesday, 27 March 2013	Min Date=2013-02-01, Max Date=2013-02-28, Rowcount=28
Thursday, 28 March 2013	Min Date=2013-02-01, Max Date=2013-02-28, Rowcount=28
Friday, 29 March 2013	Min Date=2013-02-01, Max Date=2013-02-28, Rowcount=28
Saturday, 30 March 2013	Min Date=2013-02-01, Max Date=2013-02-28, Rowcount=28
Sunday, 31 March 2013	Min Date=2013-02-01, Max Date=2013-02-28, Rowcount=28

Figure 9-4. *Shows five rows from a filtered Date Table using a Calculated Column from Listing 9-3*

In each row, the PREVIOUSMONTH function gets passed a single value from the table of dates. For the top row in Figure 9-4, the function is effectively passed a value of 26 March 2013, while in the bottom row, a single value of 31 March 2013 gets passed to the PREVIOUSMONTH function.

In the example from Listing 9-3, the PREVIOUSMONTH function always outputs a table expression that has 28 rows which get assigned to the *MyFunct* variable. The oldest date in the table expression is 1 February 2013, and the newest date is 28 February.

The text generated by the Calculated Column is useful to describe the dates inside the table expression produced by the given Time Intelligence function, based on the input. The PREVIOUSMONTH function can get replaced with another Time Intelligence function in the Calculated Column to provide a way to debug and understand what dates get considered by each function.

Now consider the following two calculated measures (or Calculated Columns) added to the same data model using the PREVIOUSMONTH Time Intelligence function.

Listing 9-4. Two calculations to help highlight the effect of the PREVIOUSMONTH function

```
Sum of Quantity = SUM('Fact Sale'[Quantity])

Sum of Quantity Prev Month =
    CALCULATE(
        [Sum of Quantity],
        PREVIOUSMONTH('Dimension Date'[Date])
        )
```

Using the calculated measures from Listing 9-4 in a Matrix visual expanded to individual day, we see the effect the PREVIOUSMONTH function has on the final values.

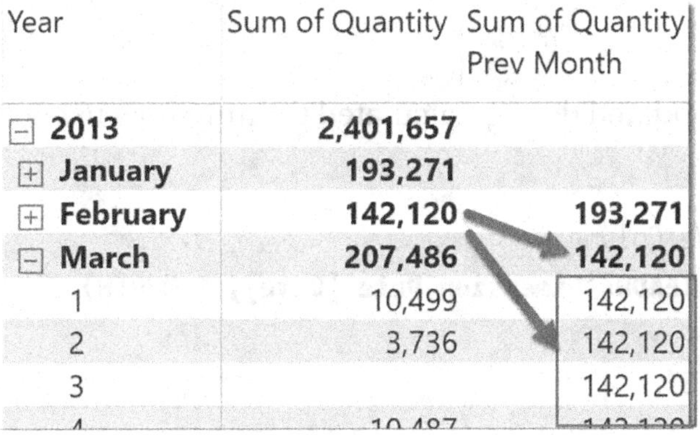

Year	Sum of Quantity	Sum of Quantity Prev Month
⊟ 2013	2,401,657	
⊞ January	193,271	
⊞ February	142,120	193,271
⊟ March	207,486	142,120
1	10,499	142,120
2	3,736	142,120
3		142,120
4	10,487	142,120

Figure 9-5. *Shows measures from Listing 9-4 in a Matrix visual*

The right-hand column in Figure 9-5 shows the rows expanded to individual day still carry a figure for the entire previous month. This result is because the table expression generated by the PREVIOUSMONTH function returns every day from the prior month and doesn't attempt to filter down to a single day to match the current day for the row. If this was the requirement, a different Time Intelligence function should be used.

The PREVIOUSMONTH function could also be called ENTIREPREVIOUSMONTH based on the output it generates.

Something else to note is the blank value in the [Sum of Quantity] column for 3 March 2013 still returns a value in the right-hand column. The blank value indicates there are no records in the Fact Sale table with an Invoice Date for 3 March 2013.

This blank cell highlights the value of having a Date Table that conforms to the principles of a Date Table. Despite no data existing in the measure table for 3 March 2013, the Date Table never has gaps, so it can still pass a value to the PREVIOUSMONTH function to derive a result for this row.

Now consider a slightly modified requirement which is to show a value that represents a value for the same day from the previous month. The PREVIOUSMONTH function returns every date from the previous month, but this requirement wants to target a specific day.

One Time Intelligence function to try is the DATEADD function. To test the behavior of DATEADD, let's update the debug column used in Listing 9-3 to use DATEADD in place of PREVIOUSMONTH.

The new debug Calculated Column gets shown in Listing 9-6. The only change made to this calculation from the same code in Listing 9-3 is the Calculated Column now uses the DATEADD function to assign a value to the *MyFunct* variable instead of PREVIOUSMONTH.

Listing 9-5. An updated debug Calculated Column now using the DATEADD function

```
TI Function Debug Column =
VAR MyFunct = DATEADD('Dimension Date'[Date],-1,MONTH)
RETURN
    COMBINEVALUES(
        ", ",
        "Min Date=" & FORMAT(MINX(MyFunct,[Date]),"YYYY-MM-dd"),
        "Max Date=" & FORMAT(MAXX(MyFunct,[Date]),"YYYY-MM-dd"),
        "Rowcount=" & COUNTROWS(MyFunct)
    )
```

A sample of the results for the updated code from Listing 9-5 gets shown in Figure 9-6. The results provide insight into the data inside the table expression returned from the DATEADD Time Intelligence function. In every case, the Rowcount shown now only has a value of 1. The Min and Max Dates are therefore always equal to each other.

Also note how in the bottom four rows the DATEADD Time Intelligence function handles the overlapping days in March 2013 that do not exist in February 2013. February 2013 only has 28 days, while Match 2013 has 31. So, for the days of the month March that are higher than the maximum date in February, the Time Intelligence function returns a table expression aligned to the last day of February.

Date ⏷	TI Function Debug Column ⏷
Tuesday, 26 March 2013	Min Date=2013-02-26, Max Date=2013-02-26, Rowcount=1
Wednesday, 27 March 2013	Min Date=2013-02-27, Max Date=2013-02-27, Rowcount=1
Thursday, 28 March 2013	Min Date=2013-02-28, Max Date=2013-02-28, Rowcount=1
Friday, 29 March 2013	Min Date=2013-02-28, Max Date=2013-02-28, Rowcount=1
Saturday, 30 March 2013	Min Date=2013-02-28, Max Date=2013-02-28, Rowcount=1
Sunday, 31 March 2013	Min Date=2013-02-28, Max Date=2013-02-28, Rowcount=1

Figure 9-6. *Shows output from the Calculated Column in Listing 9-5 on a filtered set of rows in the Dimension Date Table*

The [Sum of Quantity Prev Month] measure in Listing 9-6 uses the code from Listing 9-4 and is updated to use the DATEADD function in place of the PREVIOUSMONTH function. The updated results for the expanded Matrix get shown in Figure 9-7.

Listing 9-6. An updated calculated measure now using the DATEADD function

```
Sum of Quantity Prev Month =
    CALCULATE(
        [Sum of Quantity],
        DATEADD('Dimension Date (2)'[Date],-1,MONTH)
        )
```

Year	Sum of Quantity	Sum of Quantity Prev Month
⊟ **2013**	**2,401,657**	**2,208,196**
⊟ **2013**		
⊞ **January**	**193,271**	
⊞ **February**	**142,120**	**193,271**
⊟ **March**	**207,486**	**142,120**
1	10,499	9,934
2	3,736	4,551
3		
4	10,487	8,025

Figure 9-7. *Shows output from the Calculated Column in Listing 9-6 on a filtered set of rows in the Dimension Date Table*

The values shown in the right-hand column for the rows expanded down to individual day now are much smaller numbers. These values represent the same day from the previous month rather than for the entire previous month.

In the row that represents the entire month of March, the DATEADD returns the correct value of 142,120 for the whole of February, and even though some individual dates get duplicated at the end of the month in the level showing days (March 29 through 31), the internals inside the DATEADD function resolve this to return an accurate figure for the monthly figure.

The equivalent DAX without using a Time Intelligence function and a Date Table would be far more complicated and error-prone.

Debugging using calculated measures

The same debugging code used to create a Calculated Column can also form the basis of a calculated measure to describe data in a table expression returned by any given Time Intelligence function.

Consider the following calculation in Listing 9-7 that creates a calculated measure called [DATEADD -1]. When this measure gets added to a Matrix visual that uses a date hierarchy from the 'Dimension Date' table, the results are as displayed in Figure 9-8.

Listing 9-7. A calculated measure using the table expression produced by the DATEADD function

```
DATEADD -1 =
VAR MyFunct = DATEADD('Dimension Date'[Date],-1,MONTH)
RETURN
    COMBINEVALUES(
        ", ",
        "Min Date=" & FORMAT(MINX(MyFunct,[Date]),"YYYY-MM-dd"),
        "Max Date=" & FORMAT(MAXX(MyFunct,[Date]),"YYYY-MM-dd"),
        "Rowcount=" & COUNTROWS(MyFunct)
    )
```

Year	DATEADD -1 ▲	PREVMONTH	ENDOFMONTH
⊟ 2013			
⊟ March	Min Date=2013-02-01, Max Date=2013-02-28, Rowcount=28	Min Date=2013-02-01, Max Date=2013-02-28, Rowcount=28	Min Date=2013-03-31, Max Date=2013-03-31, Rowcount=1
1	Min Date=2013-02-01, Max Date=2013-02-01, Rowcount=1	Min Date=2013-02-01, Max Date=2013-02-28, Rowcount=28	Min Date=2013-03-31, Max Date=2013-03-31, Rowcount=1
2	Min Date=2013-02-02, Max Date=2013-02-02, Rowcount=1	Min Date=2013-02-01, Max Date=2013-02-28, Rowcount=28	Min Date=2013-03-31, Max Date=2013-03-31, Rowcount=1
3	Min Date=2013-02-03, Max Date=2013-02-03, Rowcount=1	Min Date=2013-02-01, Max Date=2013-02-28, Rowcount=28	Min Date=2013-03-31, Max Date=2013-03-31, Rowcount=1

Figure 9-8. *Shows output from three similar measures in a Matrix visual*

Two additional calculated measures get shown in the Matrix in Figure 9-8. These additional measures use the same code in Listing 9-7 but are adjusted to use the same Time Intelligence function as their name.

The first two lines for these additional measures get shown in Listing 9-8, and the results shown in Figure 9-8 describe in text the contents of the table expression stored in each calculation in the *MyFunct* variable for each day or month.

The ENDOFMONTH measure always shows a Rowcount of 1, and you can tell from the Min and Max Date values precisely what each date is for the current context.

Listing 9-8. Initial snippets for two additional calculated measures shown in Figure 9-8

```
PREVMONTH =
VAR MyFunct = PREVIOUSMONTH('Dimension Date'[Date])
<... same code as another measure...>

ENDOFMONTH =
VAR MyFunct = ENDOFMONTH('Dimension Date'[Date])
<... same code as another measure...>
```

The value of a calculated measure over a Calculated Column approach is it takes into account filter context other measures have when used in the same visual.

It can be hard to remember what similar sounding functions like PERVIOUSYEAR and SAMEPERIODLASTYEAR do concerning filtering, and debugging using this technique can help to clarify the purpose each function has.

The technique is particularly useful with functions such as DATESMTD and DATESQTD, where you see a different result for each date used in the axis.

Debugging using calculated tables

Another similar method to help understand what a Time Intelligence function returns is to use the table expression output of each function as a new calculated table. The output of the function is then available for review in the Data view screen. This approach does not take into account filter context active in visuals but does provide a quick-fire way to see what each function returns.

Consider the following two calculated *tables* in Listing 9-9.

Listing 9-9. Calculations that use Time Intelligence functions as calculated tables

```
DATESBETWEEN as Table =
    DATESBETWEEN(
        'Dimension Date'[Date],
        DATE(2013,3,3),
        DATE(2013,3,9)
        )

DATESINPERIOD as Table =
    DATESINPERIOD(
        'Dimension Date'[Date],
        DATE(2013,3,3),
        7,
        DAY
        )
```

Both examples shown in Listing 9-9 produce identical tables. The Time Intelligence functions in Listing 9-9 use slightly different syntax but output the same result. These functions are more likely to be used with dynamic variables that take into account filter context inside a measure; however, both return a single-column table with as many rows needed to satisfy the values passed as arguments. The full results of the code at Listing 9-9 are shown in Figure 9-9.

Figure 9-9. *The output from calculated tables from Listing 9-9*

Other Time Intelligence functions can be used in the same way and help validate, using tables, your understanding of the output each function produces.

Primitive vs. composite functions

Of the approximately 35 Time Intelligence functions, a handful get considered as composite functions in that they are always internally rewritten using other Time Intelligence functions.

The twelve composite functions are

DatesMTD

DatesQTD

DatesYTD

TotalMTD

TotalQTD

TotalYTD

OpeningBalanceMonth

OpeningBalanceQuarter

OpeningBalanceYear

ClosingBalanceMonth

ClosingBalanceQuarter

ClosingBalanceYear

Composite functions are made up of other Time Intelligence functions. For example, the TOTALMTD function gets converted by the DAX engine to the bottom section of code shown in Listing 9-10.

Listing 9-10. Shows composite function TOTALMTD including what it gets converted to by the DAX engine

```
TOTALMTD (
    SUM ('Fact Sale'[Quantity]),
    'Dimension Date'[Date]
    )

// Gets converted to :

CALCULATE(
    SUM ('Fact Sale'[Quantity]),
    DATESBETWEEN(
        'Dimension Date'[Date],
        STARTOFMONTH(
            LASTDATE('Dimension Date'[Date])
            ),
        LASTDATE('Dimension Date'[Date])
        )
    )
```

Both expressions return the same result as well as identical logical and physical query plans.

Figure 9-10 shows the logical plan for both DAX expressions in Listing 9-10.

The Sum VertiPaq operator at step 1 runs over the 'Fact Sale'[Quantity column] as expected, and there is a sibling operator node DatesBetween at step 2 that drops down through its children to other Time Intelligence functions such as StartOfMonth and LastDate. The structure of the logical plan closely resembles the DAX code in Listing 9-10 after conversion.

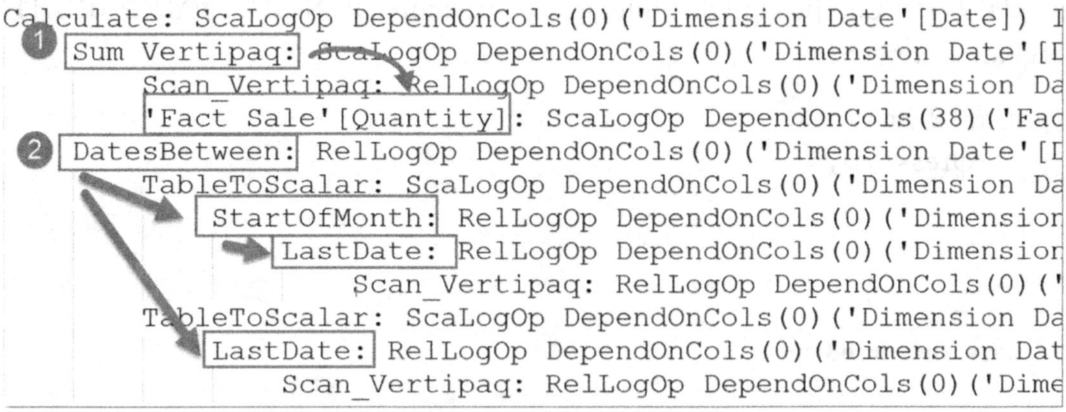

```
Calculate: ScaLogOp DependOnCols(0)('Dimension Date'[Date])  I
   Sum Vertipaq: ScaLogOp DependOnCols(0)('Dimension Date'[L
      Scan_Vertipaq: RelLogOp DependOnCols(0)('Dimension Da
      'Fact Sale'[Quantity]: ScaLogOp DependOnCols(38)('Fac
   DatesBetween: RelLogOp DependOnCols(0)('Dimension Date'[L
      TableToScalar: ScaLogOp DependOnCols(0)('Dimension Da
         StartOfMonth: RelLogOp DependOnCols(0)('Dimensior
            LastDate: RelLogOp DependOnCols(0)('Dimensior
               Scan_Vertipaq: RelLogOp DependOnCols(0)('
      TableToScalar: ScaLogOp DependOnCols(0)('Dimension Da
         LastDate: RelLogOp DependOnCols(0)('Dimension Dat
            Scan_Vertipaq: RelLogOp DependOnCols(0)('Dime
```

Figure 9-10. *Shows excerpt of the logical plan for both DAX expressions in Listing 9-10. The output from calculated tables from Listing 9-10*

The other composite Time Intelligence functions map as follows.

Listing 9-11. Composite Time Intelligence function TotalMTD including conversion

```
TotalMTD(
    <expression>,
    <dates>,
    <filter>
    )
```

// Gets converted to :

```
Calculate(
    <expression>,
    DatesMTD(<dates>),
    <filter>
    )
```

217

Listing 9-12. Composite Time Intelligence function OPENINGBALANCEMONTH including conversion

```
OpeningBalanceMonth(
    <expression>,
    <dates>,
    <filter>
    )
```

// Gets converted to :

```
Calculate(
    <expression>,
    PreviousDay(
        StartOfMonth(<dates>)
        ),
    filter
    )
```

Listing 9-13. Composite Time Intelligence function CLOSINGBALANCEMONTH including conversion

```
ClosingBalanceMonth(
    <expression>,
    <dates>,
    <filter>
    )
```

// Gets converted to :

```
Calculate(
    <expression>,
    EndOfMonth(<dates>),
    <filter>
    )
```

The composite functions shown in Listings 9-11, 9-12, and 9-13 all relate to working with monthly periods. For the composite functions that focus on other periods (quarter and year), change any instance of the text Month to the appropriate alternative (Quarter or Year). So EndOfMonth becomes EndOfQuarter, or EndOfMonth becomes EndOfYear.

Notice the TotalMTD composite function in Listing 9-11 uses another composite function called DatesMTD.

The remaining Time Intelligence functions are considered primitive and more likely to have specific operators in the logical plan that internally deal with some of the more complicated quirks of the Gregorian calendar.

Fiscal calendars

Calendars that begin on a day other than 1 January each year are a pretty standard requirement in business models to align data to dates in a financial or fiscal year. Often a data model should have the ability to provide end users with the flexibility to choose if they use fiscal or traditional calendar years.

The <year_end_date> parameter

Some Time Intelligence functions provide a method to help align reporting metrics to calendars other than the traditional year that begins on 1 January. These Time Intelligence functions have an optional parameter offering the chance to supply a <year_end_date> value. The functions that offer this option are mostly Time Intelligence functions that work with Years and include either the specific text "YEAR" or "YTD" as part of the function name.

CLOSINGBALANCE*YEAR*

DATES*YTD*

ENDOF*YEAR*

NEXT*YEAR*

OPENINGBALANCE*YEAR*

PREVIOUS*YEAR*

STARTOF*YEAR*

TOTAL*YTD*

The SAMEPERIODLASTYEAR function does not offer this option, nor is it needed to satisfy the primary intent of this function.

Each function listed provides an optional <year_end_date> parameter that allows you to specify a string in the format of "D/M" where M represents the numeric month of a year and D is the numeric calendar day of the month. This syntax means if you would like to have your fiscal year each reset on 1 July, you should hardcode this value to "30/6" to specify that your year ends on 30 June.

The example in Listing 9-14 shows the same [Sum of Quantity] measure added to a table three times. once on its own and then twice using the TOTALYTD function.

The TOTALYTD function provides a cumulative total over the data. By default, the cumulative total always resets and starts fresh from 1 January. Only when the additional <year_end_date> parameter gets provided, the running total begins again.

Listing 9-14. DAX query showing effect of the <year_end_date> parameter on the TOTALYTD function

```
DEFINE MEASURE
    'Fact Sale'[Sum of Quantity] =
        FORMAT(SUM('Fact Sale'[Quantity]),"#,###")

EVALUATE
    ADDCOLUMNS(
        VALUES('Dimension Date'[calendar month label]),
        ---------------------------------------
        "Sum of Quantity",
        [Sum of Quantity] ,
        ---------------------------------------
        "Calender YTD" ,
        TOTALYTD(
            [Sum of Quantity],
            'Dimension Date'[Date]
            ),
        ---------------------------------------
```

```
"Fiscal YTD" ,
TOTALYTD(
        [Sum of Quantity],
        'Dimension Date'[Date],
        "31/1"
        )
    )
```

The results of the query in Listing 9-14 get shown in Figure 9-11. The first column shows month values in chronological order, and the second column shows the value of the core measure. This second column is provided to show the individual value for each month and is not a cumulative figure.

The [Calendar YTD] column shows the TOTALYTD function using the default approach of resetting each year on 1 January. This effect can be seen in the red box highlighting the value of 216,337 that matches the same value shown in the [Sum of Quantity] column. The 216,337 value is also the smallest in the column. Rows below the red box in this column increment by the relevant [Sum of Quantity] value.

The final column shows the output of the TOTALYTD function when the optional <year_end_date> gets provided. In this case, a value of "31/1" gets provided which represents 31 January each year. A red box in this column highlights the value of 182,103 for February 2013. Rows below this value increment each month by the value shown in the [Sum of Quantity] column.

Dimension Date...	[Sum of Quantity]	[Calender YTD]	[Fiscal YTD]
CY2013-Sep	190,567	1,815,430	1,622,159
CY2013-Oct	198,476	2,013,906	1,820,635
CY2013-Nov	194,290	2,208,196	2,014,925
CY2013-Dec	193,461	2,401,657	2,208,386
CY2014-Jan	216,337	216,337	2,424,723
CY2014-Feb	182,103	398,440	182,103
CY2014-Mar	196,451	594,891	378,554
CY2014-Apr	209,020	803,911	587,574
CY2014-May	239,381	1,043,292	826,955

Figure 9-11. *Shows output from the query in Listing 9-14*

The other Time Intelligence functions that accept this optional parameter operate the same way. Just supply text to define the "reset" date, and the function should work as expected.

Testing fiscal functions

A suggested method of debugging/testing calculations using the <year_end_date> parameter is to check they reset on the correct day by using in a visual along with a column that represents individual days, rather than month, quarter, or year.

The first few lines of code from Listing 9-14 can be modified to use a day-based grain by changing the first parameter of the ADDCOLUMNS function to match the code shown in Listing 9-15.

Listing 9-15. A modified DAX query based on Listing 9-14 now grouping by date rather than month

```
EVALUATE
    ADDCOLUMNS(
        --VALUES('Dimension Date'[calendar month label]),
        VALUES('Dimension Date'[date]),
        ...
```

In Listing 9-15, the original VALUES line has been commented out and a new line added that also uses the VALUES function, but this time with a day-based column for more detailed grain. The resultset is now longer but provides you with the ability to scroll through the rows to check the [Fiscal YTD] resets appropriately at each boundary.

Why do I need to debug? There are some subtleties with the syntax that aren't immediately obvious, and getting this wrong doesn't necessarily throw an error.

Suppose a value of "1/36" gets supplied as the <year_end_date> parameter. This value technically means you are suggesting the fiscal year should end on a date that doesn't exist (36 January).

When a value such as "1/36" gets passed, the DAX query still succeeds and does not throw an error. The result using this value gets shown in Figure 9-12 where you can see in the final column the new year starts on 2 January. It's not immediately clear why a <year_end_date> value of "1/36" gets interpreted in this way. Debugging using this approach helps to confirm the behavior of the function and uses the day you expect. Once the calculation gets aggregated using a grain such as a month, incorrect results are difficult to detect.

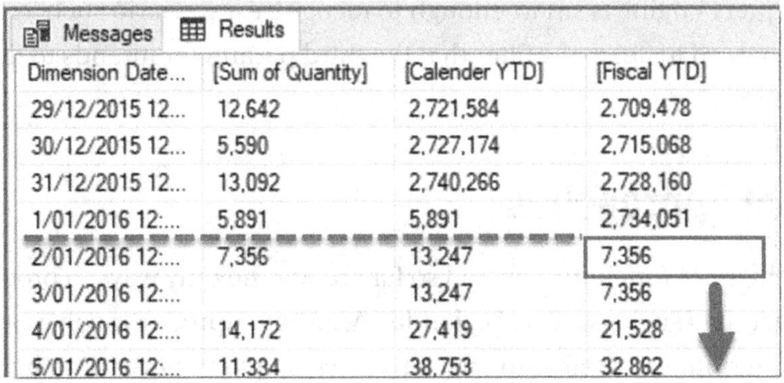

Dimension Date...	[Sum of Quantity]	[Calender YTD]	[Fiscal YTD]
29/12/2015 12...	12,642	2,721,584	2,709,478
30/12/2015 12...	5,590	2,727,174	2,715,068
31/12/2015 12...	13,092	2,740,266	2,728,160
1/01/2016 12:...	5,891	5,891	2,734,051
2/01/2016 12:...	7,356	13,247	7,356
3/01/2016 12:...		13,247	7,356
4/01/2016 12:...	14,172	27,419	21,528
5/01/2016 12:...	11,334	38,753	32,862

Figure 9-12. *Shows a sample output of query in Listing 9-14 using the <year_end_date> of "1/36"*

The DAX query engine accepts a variety of formats for the <year_end_date> parameter and attempts to resolve what it thinks you mean before resorting to an error.

"3/1" - resets 4 January

"1/3" - resets 2 February

"13/1" - resets 14 January

"1/13" - resets 14 January (same as "13/1")

"13 1" - resets 14 January

"13-1" - resets 14 January

"70-1" - resets 2 January (WTF?)

As you can see, a variety of delimiters can get used before the DAX query engine generates an error.

Another quirk of the syntax relating to the Time Intelligence functions is the position of parameters. The syntax for the TOTALYTD function according to the official documentation is

```
TOTALYTD(<expression>,<dates>[,<filter>][,<year_end_date>])
```

where the mandatory <expression> and <dates> parameters fill positions 1 and 2, respectively. The <year_end_date> parameter is actually in the fourth position, while the example code used in Listing 9-14 passes the string value to control this in position 3.

The DAX query engine is smart enough to recognize a scenario such as this when a text value is passed instead of a filter that the third parameter intends to define the <year_end_date>.

Alternative approaches

The Time Intelligence functions covered so far are not the only ways to provide alternative calendars such as a fiscal calendar. Another approach involves additional columns in a Date Table to define to which fiscal year, quarter, or month bucket each actual date should belong.

The 'Dimension Date' table in the Wide World Importers (WWI) provides a nice example to consider when designing a data model to provide this flexibility. The fiscal columns in the WWI data are

> Fiscal Year
>
> Fiscal Year Label
>
> Fiscal Month Number
>
> Fiscal Month Label

The 'Dimension Date' table in the WWI data is hardcoded to reset on 1 November each year. This pattern could be extended to include a pair of columns for Fiscal Quarter to define a text-based column for the label, for example, "FQ2014-1," with a numeric column added to manage the sorting behavior.

Often it's helpful to have a combination approach for managing fiscal calendars that uses both physical columns in Date Tables and calculations that use Time Intelligence functions where the year-end date can be defined.

It's not essential to use built-in Time Intelligence functions to manage fiscal calendars, but they are designed to make it easier to implement most time-based scenarios.

Week-based reporting

I'm a fan of a week-based grouping as I like it when metrics get aggregated and aligned to weekly buckets as it makes comparisons much more like-for-like. Weeks comprise a consistent number of days, rather than the 28–31-day mixture that can happen with months.

Much of the data we deal with has a direct or indirect relationship to the rhythm of the working week. Some retail data may peak during weekends and trough through the week. Other types of sales or utilization data may die off during the weekend, but have a consistent pattern during the week.

Comparing metrics using months may obfuscate trends when a 31-day month with five weekends is compared directly with another month with only 28 days and four weekends.

A chart plotting a metric may show sales have increased compared to the previous month, whereas a week-based axis using the same data may more correctly show a possible downward trend.

Date	Month	Week	Value
Saturday, 2 February 2019	Feb	1	200
Saturday, 9 February 2019	Feb	2	196
Saturday, 16 February 2019	Feb	3	192
Saturday, 23 February 2019	Feb	4	188
Saturday, 2 March 2019	Mar	5	184
Saturday, 9 March 2019	Mar	6	180
Saturday, 16 March 2019	Mar	7	176
Saturday, 23 March 2019	Mar	8	172
Saturday, 30 March 2019	Mar	9	168

Figure 9-13. *Shows dataset used for line charts in Figure 9-14*

Figure 9-13 shows a contrived dataset based on February and March of 2019. February has only four weekends, while March has five. The first weekend of the nine in the dataset starts with a value of 200 which then steadily decreases by precisely 4 each week. The last weekend of the nine weeks in the dataset is only 168. There is a definite and deliberate downward trend in the data.

The line chart on the left in Figure 9-14 uses calendar month on its axis and indicates an upward trend in the data, while the line chart on the right uses weeks and indicates a downward trend.

Figure 9-14. *Shows two line charts showing different trends using the same data*

The downward trend would become apparent in the line chart on the left if the user drills down through a hierarchy to a lower grain such as to an individual day, but this can't be guaranteed.

The solution is, of course, to use pie charts. Figure 9-15 shows the data from Figure 9-13 again plotted twice but this time using pie charts.

Figure 9-15. *Shows two pie charts plotting the dataset from Figure 9-13*

The two charts now no longer conflict with each other. The pie chart on the left gets divided by month and again shows a slight increase over February, while the pie chart on the right uses weeks and no longer disagrees so strongly with the trend suggested by the pie chart on the left. The only improvement that could get made to these visuals would be a 3D effect.

Only joking of course! (Joking is good for morale.) This example is the type of solution that could only be recommended by the type of project manager who also helped architect the Gregorian calendar described on the first page of this chapter. In a roundabout way, it also highlights why pie charts are a terrible choice for trend analysis. ☺

All joking aside (I wasn't serious about using pie charts in this way), week-based groupings have plenty of advantages and are useful to include in your data models. The good news is that it's not an either/or situation between weeks and months. A DAX-based data model can support the ability to slice and dice data by year/month/day as well as year/week/day. Just try not to mix them in the same visual and always use the appropriate hierarchy for your audience.

How do I use DAX to add week-based Time Intelligence to my data model? Fortunately, this is relatively easy, and DAX provides some useful date-based functions to help. These functions are

WEEKDAY

WEEKNUM

Both functions return a number and can get used to form the bases of additional DAX-based columns in any Date Table.

WEEKDAY

The WEEKDAY function accepts a date column or scalar date value and returns a number between 0 and 7 depending on the <return_type>. The <return_type> parameter can only be of values 1, 2, or 3; and the following code in Listing 9-16 shows how they differ.

SYNTAX: WEEKDAY (<date> , <return_type>)

OUTPUT: Integer

Listing 9-16. A calculated table designed to highlight the three different <return_type> options for the WEEKDAY function

```
WEEKDAY Date Table =
    ADDCOLUMNS(
        CALENDAR( DATE( 2019, 1, 1) , DATE( 2019, 1, 20) ) ,
        "Return Type 1" , WEEKDAY( [Date] , 1 ) ,
        "Return Type 2" , WEEKDAY( [Date] , 2 ) ,
        "Return Type 3" , WEEKDAY( [Date] , 3 )
    )
```

Date	Return Type 1	Return Type 2	Return Type 3
Tuesday, 1 January 2019	3	2	1
Wednesday, 2 January 2019	4	3	2
Thursday, 3 January 2019	5	4	3
Friday, 4 January 2019	6	5	4
Saturday, 5 January 2019	7	6	5
Sunday, 6 January 2019	1	7	6
Monday, 7 January 2019	2	1	0
Tuesday, 8 January 2019	3	2	1
Wednesday, 9 January 2019	4	3	2
Thursday, 10 January 2019	5	4	3
Friday, 11 January 2019	6	5	4
Saturday, 12 January 2019	7	6	5
Sunday, 13 January 2019	1	7	6
Monday, 14 January 2019	2	1	0

Figure 9-16. *Shows the Data view of the calculated table from Listing 9-16*

In Figure 9-16, in the second column [Return Type 1], the numbers range from 1 to 7, with 1 representing Sunday, 2 Monday, and so on. The first column carries a description of the weekday name.

In the third column [Return Type 2], the numbers also range from 1 to 7, but this time 1 represents Monday, 2 Tuesday, and so on.

The final column [Return Type 3] uses 0–6 as a number range with 0 representing Monday.

A few years back, I did some work for a company that liked to organize their weekly cycle by starting each week on a Saturday. It's possible to cater for this, with similar requirements by adjusting the code shown in Listing 9-16 with an offset as shown in Listing 9-17.

Listing 9-17. An adjusted calculated table using an offset to cater for weeks starting on days other than a Monday or Sunday

```
WEEKDAY Date Table =
    ADDCOLUMNS(
        CALENDAR( DATE( 2019, 1, 1) , DATE( 2019, 1, 20) ) ,
        "Return Type 2" , WEEKDAY( [Date] +2 , 2 ) ,
        "Return Type 3" , WEEKDAY( [Date] +2 , 3 )
    )
```

The updated code in Listing 9-17 provides either a 1-based number range from 1 to 7 for [Return Type 2] or a 0-based number range from 0 to 6 for [Return Type 3].

The output of this calculation can then be used to subtract days from the [Date] column to provide a [Week Starting] column in the Date Table. An example of this gets shown in Listing 9-18.

The code in Listing 9-18 shows a sample calculated table using the WEEKDAY function to generate both [Week Starting] and [Week Ending] columns for the data model. This example assumes the week begins on a Monday but can be adjusted so that weekly grouping can begin any day of the week. Figure 9-17 shows a sample of results for the code at Listing 9-18.

Listing 9-18. A method to add week-based groupings to a Date Table is using a calculated table

```
WEEKDAY Date Table =
    ADDCOLUMNS(
        CALENDAR( DATE( 2019, 1, 1) , DATE( 2019, 1, 20) ) ,
        "Week Starting" , [Date] - WEEKDAY( [Date] , 2 ) ,
        "Week Ending"   , [Date] - WEEKDAY( [Date] , 2 ) + 7
    )
```

Date	Week Starting	Week Ending
Tuesday, 1 January 2019	30/12/2018 12:00:00 AM	6/01/2019 12:00:00 AM
Wednesday, 2 January 2019	30/12/2018 12:00:00 AM	6/01/2019 12:00:00 AM
Thursday, 3 January 2019	30/12/2018 12:00:00 AM	6/01/2019 12:00:00 AM
Friday, 4 January 2019	30/12/2018 12:00:00 AM	6/01/2019 12:00:00 AM
Saturday, 5 January 2019	30/12/2018 12:00:00 AM	6/01/2019 12:00:00 AM
Sunday, 6 January 2019	30/12/2018 12:00:00 AM	6/01/2019 12:00:00 AM
Monday, 7 January 2019	6/01/2019 12:00:00 AM	13/01/2019 12:00:00 AM
Tuesday, 8 January 2019	6/01/2019 12:00:00 AM	13/01/2019 12:00:00 AM
Wednesday, 9 January 2019	6/01/2019 12:00:00 AM	13/01/2019 12:00:00 AM
Thursday, 10 January 2019	6/01/2019 12:00:00 AM	13/01/2019 12:00:00 AM
Friday, 11 January 2019	6/01/2019 12:00:00 AM	13/01/2019 12:00:00 AM
Saturday, 12 January 2019	6/01/2019 12:00:00 AM	13/01/2019 12:00:00 AM
Sunday, 13 January 2019	6/01/2019 12:00:00 AM	13/01/2019 12:00:00 AM
Monday, 14 January 2019	13/01/2019 12:00:00 AM	20/01/2019 12:00:00 AM

Figure 9-17. *A sample of output from the calculation in Listing 9-17 in the Data view*

With the [Week Starting] and [Week Ending] columns added to a Date Table, any measures using data in related child tables can get automatically grouped and sorted into 7-day buckets for easy comparison. The data type used for both columns is the Date datatype, so there is no need for additional columns to help with sorting.

WEEKNUM

Another useful DAX function to consider when working with weekly buckets is the WEEKNUM function. This function has the same parameter signature as WEEKDAY and also returns an integer value between 1 and 53. When the function returns 1, the date gets considered as being in the first week of the year.

SYNTAX: WEEKNUM (<date> , <return_type>)

OUTPUT: Integer

The various outputs of the WEEKNUM function are demonstrated using the calculated table in Listing 9-19.

Listing 9-19. A calculated table built using different <return_types>

```
WEEKNUM Date Table =
    ADDCOLUMNS(
        CALENDAR( DATE( 2019, 12, 25) , DATE( 2020, 1, 15) )
        , "Return Type 1"  , WEEKNUM([Date],    1)
        , "Return Type 2"  , WEEKNUM([Date],    2)
        --------------------------------------------
        , "Return Type 21" , WEEKNUM([Date],   21)
        , "Return Type 11" , WEEKNUM([Date],   11)
        , "Return Type 12" , WEEKNUM([Date],   12)
        , "Return Type 13" , WEEKNUM([Date],   13)
        , "Return Type 14" , WEEKNUM([Date],   14)
        , "Return Type 15" , WEEKNUM([Date],   15)
        , "Return Type 16" , WEEKNUM([Date],   16)
        , "Return Type 17" , WEEKNUM([Date],   17)
    )
```

A sample output of the code shown in Listing 9-19 can be seen in Figure 9-19. Red bars have been added to show the range for the start and end of the first week.

Only <return_types> 1 and 2 seem to be offered by the DAX intellisense in the Power BI Desktop DAX editor (January 2019 version); however, additional values for <return_type> are accepted and alternatives provided that include the ability to define weeks to start on different days.

An important aspect to highlight for the <return_types> parameter is most options cut week 1 off to always start on 1 January.

I like <return_type> = 21 that still returns a 7-day bucket for week 1 even though that means some of the earlier days in the bucket technically fall in the previous calendar year.

When working in weeks, it's probably more critical to ensure comparisons are made using the same number of days, than to try and align a weekly period with the start of the calendar year. Most of the columns in Figure 9-18 show a shortened week 1.

Date	Return Type 1	Return Type 2	Return Type 21	Return Type 11	Return Type 12
Wednesday, 25 December 2019	52	52	52	52	52
Thursday, 26 December 2019	52	52	52	52	52
Friday, 27 December 2019	52	52	52	52	52
Saturday, 28 December 2019	52	52	52	52	52
Sunday, 29 December 2019	53	52	52	52	52
Monday, 30 December 2019	53	53	1	53	52
Tuesday, 31 December 2019	53	53	1	53	53
Wednesday, 1 January 2020	1	1	1	1	1
Thursday, 2 January 2020	1	1	1	1	1
Friday, 3 January 2020	1	1	1	1	1
Saturday, 4 January 2020	1	1	1	1	1
Sunday, 5 January 2020	2	1	1	1	1
Monday, 6 January 2020	2	2	2	2	1
Tuesday, 7 January 2020	2	2	2	2	2

Figure 9-18. *Shows a sample of the output generated by code in Listing 9-19*

The WEEKNUM function is excellent for helping to build columns in a Date Table that can be used to compare weeks across years. Adding a [Week Number (Year)] column helps to tidy overhanging dates at both ends.

The code in Listing 9-20 shows a suggested Calculated Column to get added to a Date Table to allow for comparing weeks across years.

Listing 9-20. A useful Calculated Column for comparing week numbers

```
Week Number = WEEKNUM( [Date],  21)
```

The default behavior of the Calculated Column in Listing 9-20 is for the week to begin on a Monday. If another day is preferred, add a +n after the [Date], so if you prefer a Sunday the adjustment would be WEEKNUM([Date] + 1, 21).

The other Calculated Column to add to the Date Table helps separate is in Listing 9-21 and assigns each [Week Number] value into an appropriate year without losing any of the dates that overhang.

Listing 9-21. Companion column for the code in Listing 9-20 to help separate week numbers into appropriate years

```
Week Number (Year) =
VAR MyWeekNumber = 'Dimension Date'[Week Number]
VAR MyDate = 'Dimension Date'[Date]
VAR MyDates =
    FILTER(
        'Dimension Date',
        [Date] > MyDate - 6 &&
        [Date] < MyDate + 6 &&
        [Week Number] = MyWeekNumber
        )
RETURN
    IF(
        MyWeekNumber = 1,
        MAXX(MyDates,[Calendar Year]),
        MINX(MyDates,[Calendar Year])
        )
```

When the [Week Number (year)] column gets used in the legend and [Week number] column gets used on the axis of a line chart, it's possible to show metrics using like-for-like 7-day buckets across years that share the same weekday start and weekday end.

When comparing weekly buckets across years, it's often more important to make sure the buckets start/finish using the same weekdays, rather than to try and align 7-day periods using calendar days. So you probably don't want to compare the seven-day period that begins on 10 March 2019 with a seven-day period that begins on 10 March from other years.

Figure 9-19 shows a line chart with the [Week Number] column on the x-axis that uses the [Week Number (year)] column on the legend. Each data point on the line chart can be compared vertically between years to help understand the change between each year.

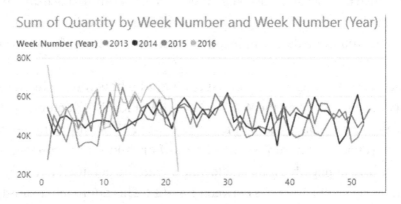

Figure 9-19. *Shows the line chart using WEEKNUM across years*

Summary

DAX is a powerful language and is more than capable of handling a wide variety of date scenarios and requirements. The data modeling engine takes advantage of more than 20 years of development, field usage, and feedback from a wide variety of users and organizations.

DAX can be used to create dynamic measures to help find insights in your data as well as to create tables to help group and organize your data into easy-to-compare buckets that make sense.

DAX can be used to create Date Tables for your data model, but this does not mean you should always use DAX to create Date Tables. Date Tables built in Power Query or from sources upstream from Power BI Desktop are equally valid and sometimes better as they may include columns already heavily customized and consistent to your organization.

A quick Internet search provides many suggestions on how to build DAX-based Date Tables. Any pattern provided by Marco Russo and Alberto Ferrari at SQLBI is an excellent starting point.

The Time Intelligence functions go a long way to help resolve the various quirks caused by our less than helpful Gregorian calendar and, when used well, reduce the amount of coding and time required to build useful reports.

This chapter has shown that DAX Time Intelligence functions typically return tables of dates, and these tables can get used by other DAX functions that accept tables as filters. The Time Intelligence tables provide appropriate unblocking semantics and have sophisticated internal code that significantly reduces the amount of code required by you.

Also covered in this chapter are some suggestions on how you might manage week-based time buckets using the WEEKDAY and WEEKNUM functions, including suggestions on how to debug. The WEEKDAY and WEEKNUM functions also provide the basis for adding dynamic columns such as [Weeks from Today] that count each weekly bucket as a positive or negative integer allowing a report to get filtered to only show values where [Weeks from Today] is between 1 and 13. This approach means the report will always so the most recent 13 weeks without interaction from end users.

Finding What's Not There

Introduction

One of the most apparent domains where DAX is applied is the calculation of measures on top of existing data, for example, calculating measures that return cumulated values over time. An area that's maybe not the first thought if one considers using DAX is to discover missing things. "Missing things" – admittedly – sounds vague; for this reason, it has to be explained in more detail. One of these missing things, for example, is a customer who stops buying.

This chapter is about finding these missing things and starts with an easy introduction about identifying customers who stopped buying. Then it's about identifying gaps between events, counting the number of gaps, and determining the length of each gap.

Finally, a distinction is made between calculating the previous value (here basically nothing is missing, and for this, this is the easy part) and finding the last row, and all of a sudden, there is a missing index.

The waning and waxing moon

The question behind this is: "Which customers bought in the last period and did not buy once again in the current period?" This will become much clearer if this will be visualized. For this reason, Figure 10-1 shows the underlying base case, and the following figures show how this question can be answered.

P. Seamark and T. Martens, *Pro DAX with Power BI*, https://doi.org/10.1007/978-1-4842-4897-3_10

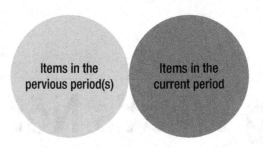

Figure 10-1. *Non-returning customer base case*

The left circle in Figure 10-1 visualizes the customers who had bought in the past. And the right circle visualizes the customers who have been buying in the current period. The conceptual answer to the question "Which customers did not buy in the current period, but did buy in the past period?" is shown in Figure 10-2.

Figure 10-2. *Non-returning customers (waning moon)*

To make things easier, let's call the shape of the circle that represents the customers who do not buy in the current period waning moon. This means that there have been more customers (items) in the past than there are today. The opposite question "Which customers just bought in the current period, but did not in the previous period?" can also be visualized in the same way. Figure 10-3 shows a waxing moon or, in a way simpler, new customers, customers who bought in the current period, but did not in the previous one.

Figure 10-3. *New customers (waxing moon)*

What has to be done to answer both of the preceding questions is to identify both sets of customers:

- Customers in the previous period(s)

- Customers in the current period

Depending on how both sets are combined, it's possible to identify the non-returning customer and also to identify the new customer. The next section shows how to answer both of the questions.

Note   This section refers to the pbix file "CH 10 Waning and Waxing moon."

New customers: Waxing moon

This chapter shows how to count customers who have not been buying in the current month in comparison to the previous month.

To determine the number of customers who did not buy in the past periods, the following measure is used (see Listing 10-1).

Listing 10-1. No of New Customer

```
No of New Customer =
var theseCustomers = VALUES('Fact Sale'[Customer Key])
var theMonthBefore = SELECTEDVALUE('Dimension Date'[RunningMonthIndex]) - 1
```

```
var theseCustomerBeforeCurrentMonth =
    CALCULATETABLE(VALUES('Fact Sale'[Customer Key]),FILTER(ALL('Dimension
    Date'[RunningMonthIndex]),'Dimension Date'[RunningMonthIndex] =
    theMonthBefore ),ALL('Dimension Date'))
return
COUNTROWS(EXCEPT(theseCustomers,theseCustomerBeforeCurrentMonth))
```

The idea behind this measure is quite simple: Just find all the customers who bought in all the periods before the current period. These customers are stored to the table variable theseCustomerBeforeCurrentMonth. All the customers who bought in the current period are stored in the variable theseCustomers. Using EXCEPT, it's in a way simple to find all the customers who did not buy in all prior periods.

Missing customers: Waning moon

To find the number of customers who did not buy in the current period, the same measure that has been used to find the number of new customers can be used. There is just a little twist. The order of tables passed to the EXCEPT function is reversed.

Each or at least: A measurement of consistency

Often consistency over time is considered a good thing, especially if positive experiences can be tied to this time. Detecting the absence of consistency or, even better, detecting when consistency is beginning to stop helps to avoid making negative experiences.

For simplicity, consistency is defined as the following: In the last 6 months, a customer must have bought anything at least four times. Consistency is endangered if the customer did not buy anything in the last two months. Figure 10-4 shows a map of various states of consistency.

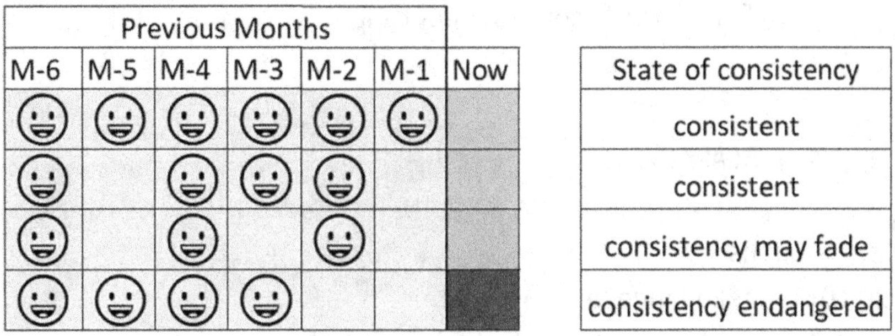

Previous Months								State of consistency
M-6	M-5	M-4	M-3	M-2	M-1	Now		
😃	😃	😃	😃	😃	😃			consistent
😃			😃	😃	😃			consistent
😃			😃		😃			consistency may fade
😃	😃	😃	😃					consistency endangered

Figure 10-4. *States of consistency*

Listing 10-2 shows the measure State of consistency.

Listing 10-2. State of consistency

```
State of Consistency =
var DateMax = MAX('Dimension Date'[Date])
// end of previous month
var DateEndOfPrevMonth =
    EOMONTH(
        DATE(YEAR(DateMax) , MONTH(DateMax) - 1, 1) ,0)
// previous 6 months
var DateStartOfRange =
    DATE(YEAR(DateMax) , MONTH(DateMax) - 6, 1)
var filtertableconsistencyrange =
    DATESBETWEEN(
        'Dimension Date'[Date]
        , DateStartOfRange
        , DateEndOfPrevMonth
    )
var consistencymayfade =
    IF(
        COUNTROWS(
            CALCULATETABLE(
                SUMMARIZE(
                    'Fact Sale'
                    , 'Dimension Date'[Calendar Month Number]
                )
```

```
                  , filtertableconsistencyrange
              )
          )
          < 4 , 1 , BLANK()
      )
// previous 2 months
var DateStartOfRangeEndangerd =
    DATE(YEAR(DateMax) , MONTH(DateMax) - 2, 1)
var filtertableconsistencyendangered =
    DATESBETWEEN(
        'Dimension Date'[Date]
        , DateStartOfRangeEndangerd
        , DateEndOfPrevMonth
    )
var consistencyendangerd =
    IF(
        COUNTROWS(
            CALCULATETABLE(
                SUMMARIZE(
                    'Fact Sale'
                    , 'Dimension Date'[Calendar Month Number]
                )
                , filtertableconsistencyendangered
            )
        )
        < 2 , 1 , BLANK()
    )
return
IF(consistencyendangerd = 1 , 3 , IF(consistencymayfade = 2 , 2 , 1))
```

What happens is this:

- A filter table filtertableconsistencyrange is determined that contains all the days of the last 6 months.

- If the number of months available in the table Fact Sale filtered by the filter table is less than 4, the value 1 is returned and stored in the variable consistencymayfade.

- A filter table `consistencyendangerd` is determined that contains all the days of the last 2 months.

- If the number of months available in the table `Fact Sale` filtered by the filter table is less than 2, the value 1 is returned and stored in the variable `consistencyendangerd.`

- Finally, a numeric value is returned: 3 represents a "consistency endangered" status, 2 represents the status "may fade," and 1 represents the status "consistent."

The measure State of consistency is used inside the measure State of consistency – Customer that is shown in Listing 10-3.

Listing 10-3. State of consistency – Customer

```
State of Consistency - Customer =
MAXX(
    VALUES('Dimension Customer'[Customer Key])
    , [State of Consistency]
)
```

The measure State of Consistency iterates over customer and returns the max value of the measure State of Consistency. This allows to "bubble up" the state if the measure is used within a hierarchical view like the Matrix visual, shown in Figure 10-5. The Matrix visual shown in Figure 10-5 can be found on the report page "Did not buy in each of the last preceding months."

Calendar Year...	Calendar Month Number	State of Consistency - Customer
☐ CY2013		
☐ CY2014	**1**	**3**
■ CY2015	Tailspin Toys (Alstead, NH)	3
☐ CY2016	Tailspin Toys (Armstrong Creek, WI)	3
	Tailspin Toys (Bowlus, MN)	3
	Tailspin Toys (Boyden Arbor, SC)	3
	Tailspin Toys (Clewiston, FL)	3
	Tailspin Toys (College Place, WA)	3
	Tailspin Toys (Crary, ND)	3
	Tailspin Toys (Cundiyo, NM)	3
	Tailspin Toys (Donner, LA)	3
	Tailspin Toys (East Fultonham, OH)	3
	Tailspin Toys (Eden Valley, MN)	3
	Tailspin Toys (Hahira, GA)	3
	Tailspin Toys (Hayes Center, NE)	3
	Tailspin Toys (Idria, CA)	3
	Tailspin Toys (Imlaystown, NJ)	3
	Tailspin Toys (Jemison, AL)	3
	Tailspin Toys (Kalvesta, KS)	3
	Total	**3**

Figure 10-5. *State of consistency – bubble up*

Measures that are just simply counting can provide tremendous insight if they are used in a little more creative way, for example, as deriving a state based on counted rows.

Sequence or the absence of events

Under certain circumstances, the knowledge of missing events is considered valuable information. Sometimes this information is much more valuable than the information about what has been measured, meaning data is available. A missing event, for example, can be considered a missing order. For this reason, not just the available information has to be analyzed, but also the events that did not occur. Knowledge about the duration of the inactivity of a customer can lead to action that prevents churn.

Here just date related sequences are considered, and as sequence maths is often much more simple if these sequences are represented by integer values instead of date or datetime datatypes, these integer values will be added to the calendar table. As the calendar table starts with 1 January 2013 and ends with 31 December 2016, an example for the preceding text will mean to create an integer for the months. This means that all the dates of January 2013 will be indexed with the number 1, all the dates of January 2014 with the number 13, and finally all the dates of December 2016 with the value 48.

Listing 10-4 shows the DAX statement that creates a Calculated Column inside the Dimension Date table.

Listing 10-4. RunningMonthIndex

```
RunningMonthIndex =
var minYear = YEAR(MIN('Dimension Date'[Date]))
var thisYear = YEAR('Dimension Date'[Date])
return
(thisYear - minYear) * 12 + MONTH('Dimension Date'[Date])
```

What happens inside the DAX statement is basically this:

- Determine the starting date inside the calendar table using YEAR(MIN(...)) and extract the starting year and store the value inside the variable minYear.

- Extract the date year from the date column of the current row and store this value inside the variable thisYear.

- Use simple math to calculate the index that represents the month as an integer.

Using months is for simplicity only. Depending on the type of business, the products sold, or many other reasons, considering the most appropriate time frame can be a very complex task, but also more revealing. These considerations are necessary when it's not expected that each customer or each product is sold with the same frequency.

Note   This section refers to the pbix file "CH 10 Waning and Waxing moon."

Listing 10-5 shows the measure SEQ – length of gap.

Listing 10-5. Sequence – length of gap

```
SEQ - length of gap =
var maxRunningMonthIndex =
    CALCULATE(
        MAX('Dimension Date'[RunningMonthIndex])
        , ALL('Dimension Date')
    )
var abt =
    ADDCOLUMNS(
        ADDCOLUMNS(
            SUMMARIZE(
                'Fact Sale'
                ,'Dimension Customer'[Customer Key]
                ,'Dimension Date'[RunningMonthIndex]
            )
            , "nxtMonth" ,
                var thisRunningMonth = 'Dimension Date'[RunningMonthIndex]
                var nxtRunningMonthFromFact =
                    CALCULATE(
                        FIRSTNONBLANK('Dimension Date'[RunningMonthIndex],
                        [Total quantity])
                        , ALL('Dimension Date')
                        , 'Dimension Date'[RunningMonthIndex] >
                        thisRunningMonth
                    )
                return
                IF(NOT(ISBLANK(nxtRunningMonthFromFact)) ,
                nxtRunningMonthFromFact , maxRunningMonthIndex)
        )
        ,"length of gap",
            var l = [nxtMonth]-[RunningMonthIndex] - 1
            return
            IF(l = 0 , BLANK() , l)
    )
```

```
return
SUMX(
    abt
    , [length of gap]
)
```

What happens is this: A summarized table is created with all the valid combinations of the columns Customer Key and the RunningMonthIndex, meaning existent inside the table Fact Sale.

This is done by this DAX:

```
SUMMARIZE(
    'Fact Sale'
    , 'Dimension Customer'[Customer Key]
    , 'Dimension Date'[RunningMonthIndex]
)
```

ADDCOLUMNS is used to add the column nxtMonth to the table. This column finds the next month index by using FIRSTNONBLANK. CALCULATE has to be used to transform the existing row context introduced by ADDCOLUMNS and to add the value of the column Customer Key to the new filter context. This is done by the following DAX:

```
CALCULATE(
    FIRSTNONBLANK(
            'Dimension Date'[RunningMonthIndex]
            , [Total quantity]
    )
    , ALL('Dimension Date')
    , 'Dimension Date'[RunningMonthIndex] > thisRunningMonth
)
```

The result is stored to the variable nxtRunningMonthFromFact. This step is necessary to avoid negative values. If there is no next value, in this case, the latest index will be used.

Then, with a little math, the difference between the next month and the current month is calculated; this returns the length of the gap in months.

Finally, the table is used in combination with the iterator function SUMX that creates the sum of all the gaps.

The second measure, SEQ – no of gaps, is basically doing the same with just the little difference that it counts the gaps.

Figure 10-6 shows a simple table.

Customer Key	Calendar Month Label	SEQ - no of gaps	SEQ - length of gap
390	CY2016-May	1	6
399	CY2016-May	1	6
401	CY2016-May	1	6
402	CY2016-May	1	6
123	CY2013-Dec	1	3
334	CY2014-May	1	3
182	CY2014-Oct	1	3
175	CY2015-Oct	1	3
59	CY2013-Jan	1	2
84	CY2013-Jan	1	2
148	CY2013-Jan	1	2
180	CY2013-Jan	1	2
Total		**1534**	**7**

Figure 10-6. *SEQ table*

Both measures can be considered as the start on how to analyze missing events.

The missing index

Similar but different is the quest for the previous row. The task that has to be solved is finding the previous row. Even if this may sound similar to finding the previous value, it is not. Finding the previous row in a large dataset is not an easy task. For this, it's necessary to start with explaining the difference.

DAX has been often compared to other query languages like SQL or MDX. In some of these comparisons, DAX provides better performance and in some it does not. DAX does not provide similar functions to the T-SQL window functions like ROW_NUMBER, LAG, or LEAD. These functions are very powerful and allow to number a set of rows (this is not the same as providing a rank). LAG and LEAD allow to navigate between rows, for example, find preceding or subsequent rows.

Sometimes it's necessary to calculate the previous value inside a more complex measure, for example, calculating a Year-over-Year (YoY) comparison.

Sometimes it's necessary to find the previous row, for example, to determine the change on a row-by-row basis.

Even if both tasks look similar, they are not. It's necessary to be able to differentiate between both tasks. For this reason, the next two sections will explain these tasks in more detail.

Previous value

If it's possible to determine the previous value from a given context, it is the previous value problem.

Figure 10-7 will show what this is about.

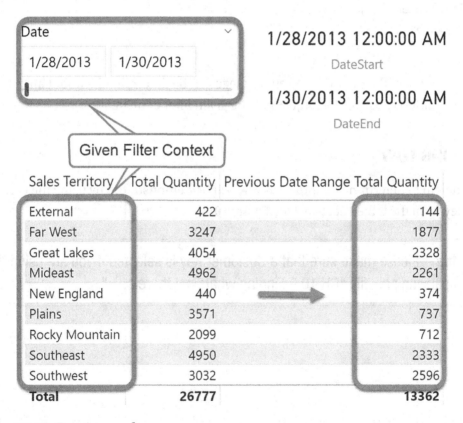

Figure 10-7. *Previous value*

Listing 10-6 shows how the range of necessary dates to filter the preceding date range can be derived from the given filter context.

Listing 10-6. Previous Date Range Total Quantity

```
Previous Date Range Total Quantity =
var DateRangeStart = [DateStart]
var DateRangeLength = DATEDIFF([DateStart] , [DateEnd] , DAY)
var PrevDateRangeEnd = DateRangeStart - 1
var PrevDateRangeStart = PrevDateRangeEnd - DateRangeLength
var PrevDateRange = DATESBETWEEN('Dimension Date'[Date]
, PrevDateRangeStart , PrevDateRangeEnd)
return
CALCULATE(
    [Total Quantity]
    , PrevDateRange
)
```

The content of the filter table `PreDateRange` can be determined without any additional table scans, just by "adjusting" the given filter context.

Previous row

The problem to determine the previous row will be demonstrated by using the column Sale Key from the table Fact Sale (see Figure 10-8).

Warning   Please make sure that a Customer Key is selected if you are using the accompanying pbix file "CH 10 – Finding whats not there.pbix".

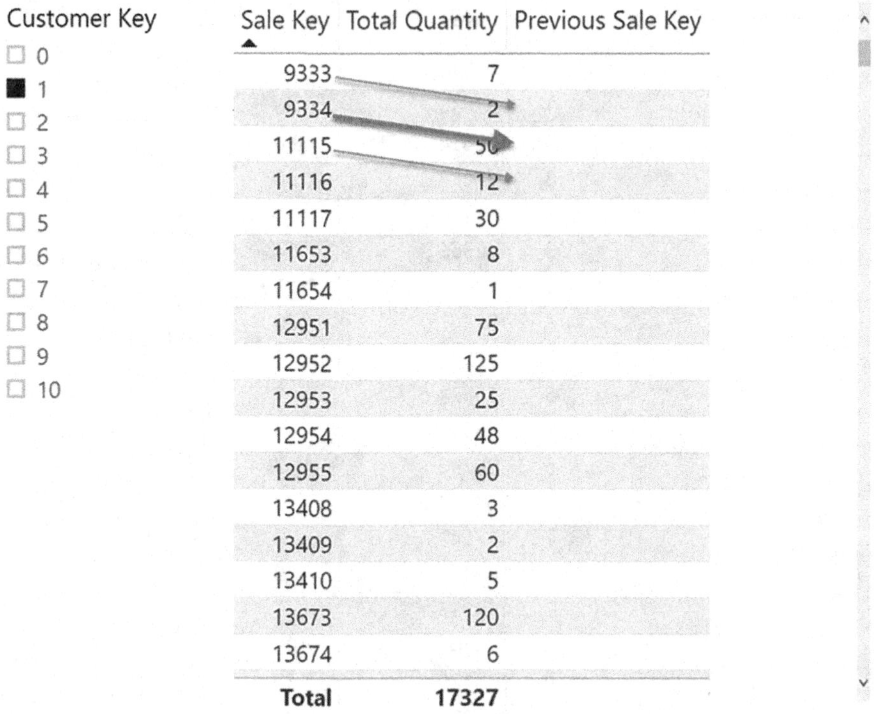

Customer Key	Sale Key	Total Quantity	Previous Sale Key
☐ 0	9333	7	
■ 1	9334	2	
☐ 2	11115	50	
☐ 3	11116	12	
☐ 4	11117	30	
☐ 5	11653	8	
☐ 6	11654	1	
☐ 7	12951	75	
☐ 8	12952	125	
☐ 9	12953	25	
☐ 10	12954	48	
	12955	60	
	13408	3	
	13409	2	
	13410	5	
	13673	120	
	13674	6	
	Total	**17327**	

Figure 10-8. *Previous row*

A naïve approach is to determine the previous row by simply subtracting 1 from the current Sale Key:

PreviousRow(SaleKey: 9334) = 9334 – 1 = 9333

The complexity becomes more evident in Figure 10-9.

Customer Key	Stock Item Key	Sale Key	Total Quantity	Previous Sale Key
☐ 0				
■ 1	1	14574	10	
☐ 2	1	127538	40	
☐ 3	2	155414	80	
☐ 4	3	11117	30	
☐ 5	3	61162	10	
☐ 6	3	142352	60	
☐ 7	3	158062	50	
☐ 8	4	182447	20	
☐ 9	5	117544	7	
☐ 10	5	125334	9	
	6	58510	3	
	6	96790	7	
	6	164023	9	
	7	117542	8	
	7	182445	9	
	7	184835	2	
	9	124507	100	
	9	133126	30	
	11	203929	35	
	12	101130	84	
	12	142350	60	
	12	144066	48	
	12	203241	84	
Total			**17327**	

Figure 10-9. *Previous row – more complex*

Figure 10-9 shows that next to the value of the column Customer Key, also the value of the column Stock Item Key has to be considered to determine the previous row.

Finding the previous row: Using DAX

To find the previous row using DAX, it is necessary to understand what steps have to be accomplished to find the previous row.

For this example, the column Sale Key from the table Fact Sale is used as the values can be used to identify every single row.

As this is not a naïve approach, the solution that is used also takes into account the Customer Key. Figure 10-10 shows a little conceptual table to better understand these steps.

Sale Key	Customer Key	Quantity	Previous Sale Key	Previous Sale Value
1	A	11	#N/A	#N/A
2	A	12	1	11
3	B	10	#N/A	#N/A
4	A	9	2	12
5	B	9	3	10

Figure 10-10. *Previous row – concept*

- Create a table to iterate over.

  Available combinations (available in the current filter context) of Customer Key and Sale Key.

- Determine the previous Sale Key.

  The previous Sale Key is the most recent Sale Key prior to the current Sale Key.

  The current Sale Key (from the current iteration) will be used to find all the Sales Keys less than the current one. The most recent Sale Key prior to the current Sale Key is the MAX('..'[Sale Key]) from all the Sale Keys smaller than the current Sale Key.

- Retrieve the previous value.

  Retrieve the previous value by using the previous Sale Key to filter the Fact Sale table.

Caution   Before you add the measure previous sale value from Listing 10-7 to the table on the report page "Previous Row" of the Power BI file "CH 10 Finding whats not there.pbix", make sure that the Customer Key slicer only selects a single Customer Key.

The next listing, Listing 10-7, shows a "DAX-only" solution to find the previous row. A mandatory requirement is the presence of a unique identifier that allows finding the previous row.

Listing 10-7. previous sale value

```
previous sale value =
var t =
    ADDCOLUMNS(
        FILTER(
            SUMMARIZE(
                'Fact Sale'
                , 'Dimension Customer'[Customer Key]
                , 'Fact Sale'[Sale Key]
            )
            , [Customer Key] <> 0 // customer key 0 (unknown) is not
            considered
        )
        , "thePrevQuantity" ,
            var currentSaleKey = [Sale Key]
            var prevsalekey =
                MAXX(
                    FILTER(
                        CALCULATETABLE(
                            VALUES('Fact Sale'[Sale Key])
                            , ALL('Fact Sale'[Sale Key])
                        )
                        , 'Fact Sale'[Sale Key] < currentSaleKey
                    )
                    , 'Fact Sale'[Sale Key]
                )
            var prevQuantity =
                CALCULATE(
                    [Total Quantity]
                    , 'Fact Sale'[Sale Key] =  prevsalekey
                )

            return prevQuantity

    )
return
```

```
SUMX(
    t
    , [thePrevQuantity]
)
```

The preceding listing will now be explained in greater detail.

First, all the available (available in the current filter context) combinations of Customer Key and Sale Key will be summarized by this DAX:

```
SUMMARIZE(
'Fact Sale'
, 'Dimension Customer'[Customer Key]
, 'Fact Sale'[Sale Key]
)
```

In addition to this, FILTER is used to exclude the Customer Key 0, as this key represents the customer "unknown."

ADDCOLUMNS adds the column thePrevQuantity to the table. It's necessary to remember that ADDCOLUMNS is one of the table iterator functions. This means that the expression used to calculate thePrevQuantity is evaluated in a row context, determined by the columns Customer Key and Sale Key.

This DAX

```
CALCULATETABLE(
    VALUES('Fact Sale'[Sale Key])
    , ALL('Fact Sale'[Sale Key])
)
```

expands the current combination of Customer Key and Sale Key to all the Sales Keys of the current Customer Key in the current filter context. This is because CALCULATETABLE does the following:

- Clone the existing filter context into a new filter context.

- Add the current row context to the new filter context (for each iteration), where the existing filter (already existing in the cloned filter context) is blocked; it is maybe helpful to visualize that the existing filter on the columns Customer Key and Sale Key is replaced by the columns of the row context.

- Evaluate the filter expression ALL(…) in the old filter context, and "replace" the already existing filter on the column Sale Key in the new filter context.

- Evaluate the first parameter VALUES('Fact Sale'[Sale Key]) in the new context.

The result of this step is visualized in Figure 10-11.

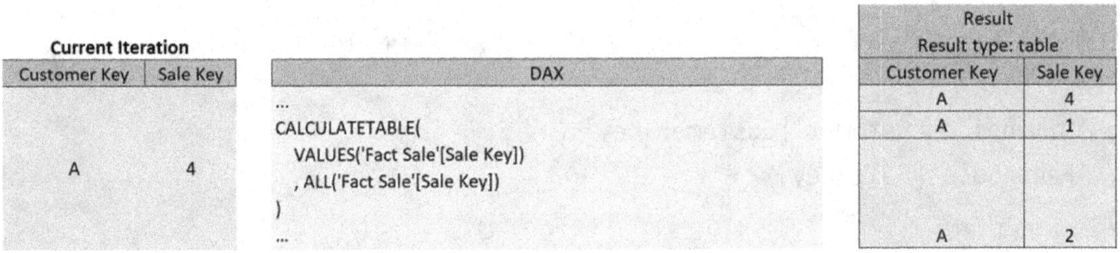

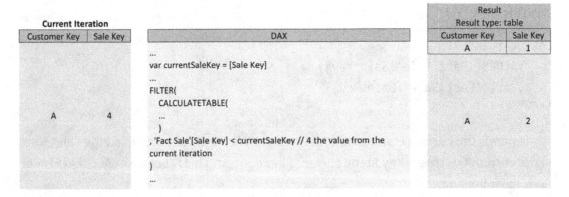

Figure 10-11. *Previous row – concept, expand*

Then the resulting table of the previous step is filtered down to all the rows where the Sale Key is less than the current Sale Key (stored to the variable currentSaleKey). The result is shown in Figure 10-12.

Figure 10-12. *Previous row – concept – filter*

MAXX is used to find the max value in the filtered table object returned by FILTER(…). This returns the previous Sale Key. The scalar value returned by MAXX is stored in the variable prevsalekey. This variable is finally used to retrieve the quantity of the previous Sale Key.

Finding the previous row: Using a modeling approach

Another approach to solve the previous row problem is a modeling-based approach by using an additional table that already contains the previous value. This table has to be created outside of the data model, for example, using Power Query or – even better – inside the source system. This approach can help to overcome performance issues with large datasets. Creating this table can become challenging but can help to overcome performance issues with large datasets. The pbix file contains the table Fact Sale – Previous Value that has been created inside SQL Server using the SQL statement from Listing 10-8. It is strongly recommended to "hide" this additional table from the end user of the data model to avoid confusion.

Listing 10-8. T-SQL using LAG to find the previous value.sql

```
select
f.[Customer Key]
, f.[Sale Key]
, f.[quantity]
, LAG(f.[Sale Key] , 1) over(partition by f.[Customer Key] order by f.[Sale
Key] asc) as prevsaleeky
, LAG(f.[Quantity] , 1) over(partition by f.[Customer Key] order by f.[Sale
Key] asc) as prevquantity
from
    Fact.Sale as f
```

This additional table then is integrated into the data model by creating a relationship between the tables Fact Sale and Fact Sale – Previous Sale using the column Sale Key. As these columns are unique in either table, Power BI creates a relationship of the cardinality type One to one (1:1). It's necessary to change the cardinality to One to many (1:*) with the table Fact Sale on the one side to fully leverage the speed of filter propagation. This relationship is shown in Figure 10-13.

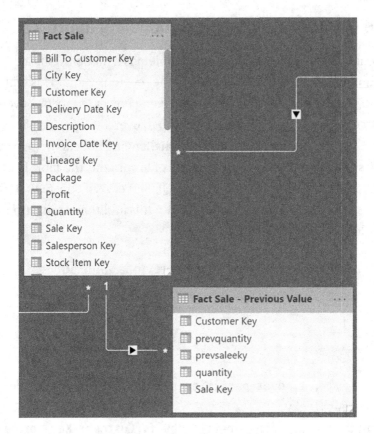

Figure 10-13. *Previous row, fact-based – relationship*

A measure as simple as that shown in Listing 10-9 can be used to calculate the previous value.

Listing 10-9. Previous sale value – fact-based

```
previous sale value - fact based =
SUM('Fact Sale - Previous Value'[prevquantity])
```

Using a modeling approach has a tremendous performance advantage. This performance advantage is shown in the next two figures. Figure 10-14 shows the query duration for calculating the previous value using the DAX-only approach for the Customer Key 2 and for the Customer Keys 2–200. Figure 10-15 shows the same but using the additional fact table Fact Sale - Previous Value.

StartTime	Type	Duration	User	Database	Query
11:38:56	DAX	7,979	tmart	CH 10 fin...	DEFINE VAR __DSOFilterTable = FILTI
11:38:28	DAX	211	tmart	CH 10 fin...	DEFINE VAR __DSOFilterTable = FILTI
11:38:28	DAX	2	tmart	CH 10 fin...	EVALUATE ROW("MinCustomer_Key

Figure 10-14. *Previous row value, DAX only*

StartTime	Type	Duration	User	Database	Query
07:02:23	DAX	76	tmart	CH 10 fin...	DEFINE VAR __DSOFilterTable = FILTER(KEEPFILTI
07:01:43	DAX	26	tmart	CH 10 fin...	DEFINE VAR __DSOFilterTable = FILTER(KEEPFILTI
07:01:43	DAX	3	tmart	CH 10 fin...	EVALUATE ROW("MinCustomer_Key", CALCULAT

Figure 10-15. *Previous row value, fact-based*

In both figures, the bottommost query is issued to fill the Customer Key slicer; the tiny difference of 1 millisecond does not count.

The middle line shows the calculation of the previous row value for Customer Key 2, even if this difference seems not big as both queries return the answer in less than one second.

But the difference between the topmost queries is tremendous. It takes **almost 8 seconds** to calculate the previous row value for the Customer Keys 2–200 using the "DAX-only" approach.

The simple measure leveraging the additional fact table returns the result in **less than 80 milliseconds.** This is a huge performance gain.

Caution   Using an additional fact table also has it downsides. As this additional table has the same granularity as the fact table, it will, of course, consume additional memory. Creating an additional fact table can also require a much more complex SQL statement than the one from the preceding text because determining the previous Sale Key can involve much more logic than the simple approach used for this example.

There is no clear answer how to tackle the previous row problem, but hopefully this demonstrates that a complex and slower DAX statement can (maybe) be simplified by using all of the underlying capabilities of the database engine: data modeling and DAX.

CHAPTER 11

Row-Level Security

Introduction

At some point, you may come across a requirement to create a report that needs to deliver results specific to the currently logged in user who looks at the report. One scenario is when different team members need a report that shows low-level detail about items they are individually responsible for, but prevents them from seeing the same level of detail for other members in the same team. Another scenario might be to provide a report to multiple groups external to your organization, making sure the person accessing the report only sees the information they are supposed to.

One solution might be to create a master copy of your data model and make as many clones as needed to satisfy the number of information silos required. Each clone can get populated with data specific to a user or organization, and then each report gets provided to just the user who should have access. This approach reduces the risk of data leakage.

However, this approach can become complicated and unwieldy to manage once the number of data silos gets to be more than a handful. When changes need to get made to calculations that already exist in the model or new calculations get added, each data model needs to be modified, tested, uploaded, and so on.

Other issues to be wary of with approach are around scheduling of data refreshes, model size, and managing the unique code in each data model that controls the import of only appropriate data.

Fortunately, there is an alternative solution in Row-Level Security (RLS). RLS allows you to bring all relevant data into a single data model and provide a secure and robust layer of filtering over the data to manage who sees what.

The concept of Row-Level Security is not new. SQL Server Analysis Services databases have been offering the concept of Row-Level Security for a long time now with both the Multidimensional and Tabular versions offering this type of functionality.

© Philip Seamark, Thomas Martens 2019
P. Seamark and T. Martens, *Pro DAX with Power BI*, https://doi.org/10.1007/978-1-4842-4897-3_11

Power BI Desktop and the Power BI web service use the Tabular version of the SSAS database, and Microsoft understands and appreciates how critical this feature is for organizations to manage their data.

With RLS configured on a single data model, different users can access the same model, but safely preventing them from seeing data they should not be able to. Some advantages of using RLS are when new calculations get added or existing calculations updated, changes only need to get made in one place. Administering a single model becomes much easier regarding data refreshing, gateways, and sizing, to mention just a few.

With RLS, rules can get defined so data can get filtered in many different ways. A supplier may be granted access to a data model to see stock and sales data for their products but not for other suppliers or entire tables in the same model, such as employee or financial data.

In multi-store environments, individuals within a specific store could access a report using a data model with RLS and drill to very low-level detail to see how their store is performing while also seeing aggregated information about other stores that provide a useful benchmark for comparison. Individuals at other stores can simultaneously access the same data model with the same restrictions applied.

Employees within your organization may need access to specific data in the model but not sensitive data (financial, payroll). Others may need access to high-level aggregated data, but be prevented from drilling down through to the lower levels of hierarchies that may expose sensitive or personally identifiable information they do not need to access, while others using the same model need full access to details where appropriate.

RLS is flexible enough to cater for any requirement you are likely to need. The limitations on rules for RLS are not necessarily technical, but more around manageability and maintainability when it comes to a large number of complex rules – especially when it comes to rules that conflict or overlap.

The good news is that the rules used to configure RLS are all DAX-based. RLS rules get defined as DAX filter expressions and can be as simple as [Country] = "USA" or as a complex, multilayered series of statements that ultimately get resolved down to a Boolean true or false to decide which rows should get filtered in or out.

One way to think of the RLS filter expressions is like a T-SQL WHERE clause that gets added to every query and every visual in a report. Another way to think of RLS is a multi-pass process. If RLS is defined, the first pass is to apply the RLS rules to a table, and only rows that survive the first RLS pass are visible to queries from the report.

Roles

RLS in Power BI is role based. At least one role must get defined in the model if you would like to use RLS. You can create as many roles as you like. Roles can be tailored to fit individual users, groups, or other non–user-specific entities. Roles can be departmental, regional, or even time-based.

Power BI Desktop allows you to create roles but does not manage who gets assigned to each role. The work of assigning users to roles is done using the Power BI web service. This activity of assigning users to roles can be performed and managed by someone entirely different from the person who configures rules for each role.

In Power BI Desktop, to add or edit roles, click the Manage Roles button in the Security section of the Modeling tab in the ribbon. An image of the Manage Roles button from the ribbon gets shown in Figure 11-1.

Figure 11-1. *Shows the Manage Roles button to access RLS roles*

Clicking the Manage Roles button brings up a dialog window that shows a list of every role currently defined in the data model.

The three elements of the Manage Roles dialog, seen in Figure 11-2, are Roles, Tables, and Filters.

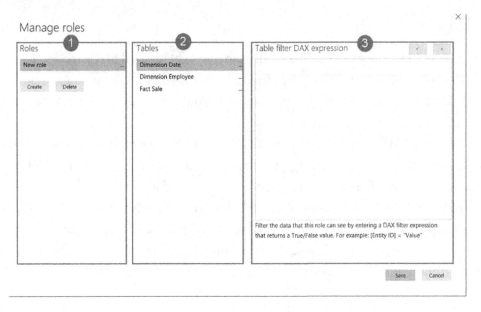

Figure 11-2. *The Manage Roles dialog window in Power BI Desktop*

(1)   Roles

(2)   Tables

(3)   Filters

Roles

The Roles section at step 1 shows a list of currently defined roles. This section provides a button to create new roles or delete existing roles. Clicking a role name in the list of roles updates the information in the Tables and Filters sections to show details specific to the highlighted role. Only one role can get highlighted at the same time.

An ellipsis at the right-hand side of the role name provides the ability to rename or clone an existing role.

Tables

Next to the Roles section is a list showing every table in the data model. The Tables panel is shown in step 2 of Figure 11-2. Clicking the ellipsis on the right-hand side, a table name opens a context menu that initially offers options to "Add Filter" or "Clear table filter."

Expanding the "Add Filter" option lists every column in the table, and clicking any column creates or extends the DAX filter expression in the Filters section.

Once a table has at least one filter defined, a filter icon appears next to the ellipsis for the table to show a filter now exists. Clicking the table name updates the text in the Filters section to show the current DAX filter expression for that table. If a column name gets clicked in the context menu, the column name appears in the Filters window with a sample operator and value to help guide appropriate syntax.

Clicking the [Date] column in the 'Dimension Date' table populates the Filters section with a suggested DAX filter expression. The suggestion takes into account the datatype of the column selected to get filtered, and different suggestions are provided for numeric and text columns. An example of this gets shown in Figure 11-3.

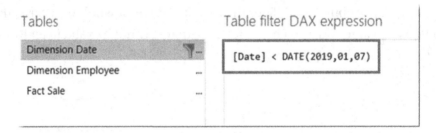

Figure 11-3. *Example of RLS DAX filter expression on the 'Dimension Date' table*

Filters

The final section of the Manage Roles dialog is the Filters section at step 3 of Figure 11-2. The Filters section provides a panel showing the current DAX expression used to filter rows for the selected table/role. The text in the window can get edited directly and, as a whole expression, must evaluate to a Boolean true or false.

Subqueries can get used in RLS filter statements, and some examples are provided later in this chapter in section "Dynamic RLS using subqueries."

Filters can get defined over multiple tables for a given role which means a single role can have a filter definition similar to the following:

Geography [Country] IN {"USA","Canada"}

Date [Calendar Year] >= 2020

Employee [Department] = "Sales"

In the preceding filter definition for a role, three tables have filters applied. Each row in the Geography table is assessed by a rule that states the value in the [Country] column in the 'Geography' table must match either "USA" or "Canada". Note the use of the DAX IN function to allow multiple selections.

The Date Table also has a rule that filters out years before 2020, and finally, the Employee table gets filtered so that only rows with "Sales" in the [Department] column get returned.

Let's assume each table in the example is a dimension table in a star schema that sits on the one side of some one-to-many relationships with some fact tables. These rules automatically propagate down through the one-to-many relationships and get applied to downstream tables. The rules do not have to be on the column that defines a relationship.

The Filters section also provides two buttons in the top right-hand corner as shown in Figure 11-4. The tick button validates the DAX in the editor window to verify that it returns a Boolean value and also checks for code that may not be valid in an RLS filter statement.

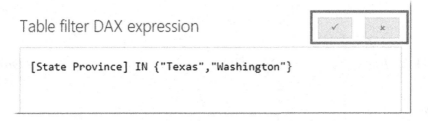

Figure 11-4. *The buttons to check or clear an RLS filter rule*

The button with the cross clears all tests in the window for the currently selected table.

Testing roles

As with any form of development, it's essential to test your changes to make sure they behave as expected. Power BI Desktop provides a helpful method to test RLS roles to help identify issues as early as possible. RLS is particularly critical if data in the model is sensitive and should never be visible to certain people.

Next to the Manage Roles button in the ribbon is the View as Roles button. Clicking this opens a dialog that lets you select the name of any role that has been defined using the Manage Roles screen. The button gets shown in Figure 11-5.

Figure 11-5. *The View as Roles button in the Security section of the Modeling tab*

Assume a single role called "Test Role 1" is defined in a data model using the 'Dimension City', 'Dimension Date', and 'Fact Sale' tables from the Wide World Importers dataset. A standard one-to-many relation exists between 'Dimension City' and 'Fact Sale' using the [City Key] column, while the 'Dimension Date' table is connected to the 'Fact Sale' table using the [Invoice Date Key] column. Figure 11-6 shows the Relationships view for the current example.

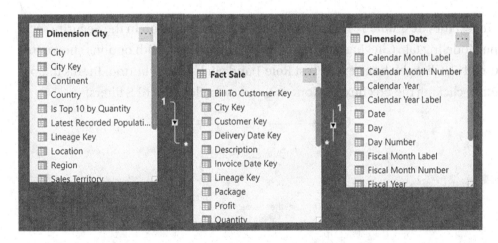

Figure 11-6. *The Relationships view of the test example*

The "Test Role 1" role has the following RLS filter rules applied:

'Dimension City' [State Province] IN {"Texas","Washington"}

'Dimension Date' [Calendar Year] = 2015

The relationship between the 'Dimension Date' and 'Fact Sale' tables uses the [Invoice Date Key] column. In this example, there is only a single relationship between these two tables, and the relationship gets configured as an active relationship.

RLS filtering rules only propagate down via the active relationship, so additional *inactive* relationships are meaningless for RLS.

To establish a baseline for the test, a [Sum of Quantity] calculated measure using the code in Listing 11-1 gets added to a blank report page with no other filters. The result of 8,950,628 gets shown in a visual in Figure 11-7.

Listing 11-1. A basic calculated measure to help test RLS

```
Sum of Quantity = SUM('Fact Sale'[Quantity])
```

8,950,628

Sum of Quantity

Figure 11-7. *The unfiltered result of the [Sum of Quantity] calculated measure*

To test the "Test Role 1" role, click the View as Roles button in the Security section of the Modeling tab. Clicking this button brings up a dialog with options shown in Figure 11-8. Click the box next to Test Role 1 and click the OK button. In this dialog window, clicking the box next to None is one way to clear the RLS filters.

View as roles

☐ None

☐ Other user

☑ Test Role 1

Figure 11-8. *The View as Roles dialog window*

The unfiltered visual using the calculated measure in Figure 11-7 now shows a different value. The value of 244,520 shown in Figure 11-9, takes into account RLS and gets filtered by the rules defined over both the 'Dimension City' and 'Dimension Date' tables. No other filters are applied directly to the 'Fact Sale' table in the report.

244,520

Sum of Quantity

Figure 11-9. *The unfiltered result of the [Sum of Quantity] calculated measure using RLS*

The Report and Data views of Power BI Desktop now show a yellow band immediately underneath the ribbon with a message that shows you are "Now viewing the report as: Test Role 1." This warning message gets shown in Figure 11-10. A button on this warning message removes the RLS filtering when clicked.

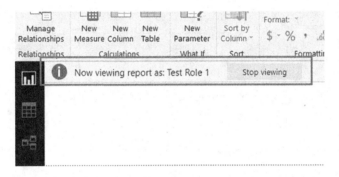

Figure 11-10. *The unfiltered result of the [Sum of Quantity] calculated measure*

Another way to review the test is via the Power BI Desktop Data view window. The Data view only shows rows that meet the RLS filter requirements for the active role. When no role gets tested, the Data view shows every row in every table. When an RLS role gets tested, the row count is updated to reflect RLS filtering. Table 11-1 shows the effect of the "Test Role 1" on the data visible in the Data view. There is no change to the 'Dimension Employee' table because this specific test applies no filtering to this table.

Table 11-1. *Effect of the RLS role on row counts of tables in the data model*

Table Name	Rowcount without RLS	Rowcount with RLS test
Dimension City	116,295	10,427
Dimension Date	1,461	365
Dimension Employee	213	213
Fact Sale	228,265	6,378

A quick check of the 'Fact Sale' table is to sort the [Invoice Date Key] column and check that the top and bottom boundary values still match the 2015 filter requirement of being in the calendar year of 2015.

RLS filtering is not just for calculated measures or visuals; filtering takes place at a higher level in the data model, so even the data visible in the Data view gets restricted.

Testing multiple roles

When dynamic RLS gets used in your data model and you have more than one role, it's possible a user of your published dataset might exist in multiple roles. To help test this in Power BI Desktop, enable multiple checkboxes in the View as Roles dialog as per Figure 11-11.

Figure 11-11. *The View as Roles dialog showing multiple roles selected*

Be aware that Power BI Desktop treats the rules across roles as an OR rather than an AND.

Consider the following scenario.

"Test Role 1" role has the following RLS filter rules defined:

'Dimension City' [State Province] = "Texas"

"Test Role 2" role has the following RLS filter rules defined.

'Dimension Date' [Calendar Year] = 2015

When only Test Role 1 is selected, data in the 'Fact Sale' table gets filtered down to only rows belonging to "Texas".

When only Test Role 2 is selected, data in the 'Fact Sale' table gets filtered to only show rows belonging to 2015.

However, when both Test Role 1 and Test Role 2 are selected, rows in 'Fact Sale' can either belong to "Texas" OR 2015. Rows belonging to other [State Provinces] appear but only for 2015.

This effect can be seen in Figure 11-12 using a Matrix visual that has [Calendar Year] on columns and [State Province] on rows. This view gets captured when both "Test Role 1" and "Test Role 2" are simultaneously selected using the View as Roles dialog.

State Province	2013	2014	2015	2016
South Dakota			35,536	
Tennessee			19,815	
Texas	170,541	199,032	183,274	74,425
Utah			20,273	
Vermont			22,035	
Virginia			30,981	
Washington			61,246	
West Virginia			58,609	
Wisconsin			55,648	
Wyoming			25,243	

Figure 11-12. *A Matrix visual highlighting how multiple roles are treated using OR rather than AND*

The result shown in Figure 11-12 is also reflected in the Power BI web service and is vital to understand when designing a data model with more than one RLS role. Remember that a user account can get assigned to more than one role in the Power BI web service.

Active relationships and RLS

Using the previous example, let's look at what happens when we change the column involved in the active relationship between the 'Dimension Date' and 'Fact Sale' tables. The current relationship links the [Invoice Date Key] column with the [Date] column in the 'Dimension Date' table.

Let's update the relationship to now use the [Delivery Date Key] column in the 'Fact Sale' table and see what happens to the Report and Data views. The RLS filtering rules defined for the "Test Role 1" role do not change.

The visual in the Report view now shows a different value from what got seen previously. The new value shown in Figure 11-13 is 243,923, and this was previously 244,520. This reflects the RLS filter now gets applied over the 'Dimension Date' table and propagates down to a different set of rows in the 'Fact Sale' table.

243,923

Sum of Quantity

Figure 11-13. *The updated value of the calculated measure after relationship change*

DAX Query Plan

Let's have a look at some DAX profiler events for the preceding example using trace events recommended in Chapter 13, "Query Plans."

The recommended trace events are

> Query Events
>
> > Query End
>
> Query Processing
>
> > DAX Query Plan
> >
> > VertiPaq SE Query Cache Match
> >
> > VertiPaq SE Query End

With these events enabled in the SSMS profiler, using a cold cache, the following events get captured when refreshing the visual shown in Figure 11-9.

Logical Plan

A simplified version of the logical plan gets shown in Listing 11-2, and there are no clues or references in the logical plan regarding any filtering rules applied by RLS. The logical plan suggests a Sum_VertiPaq operation over the 'Fact Sale'[Quantity] column.

Listing 11-2. A simplified version of the logical plan captured when testing the calculated measure using the RLS role

```
AddColumns:
    Sum_Vertipaq:
            Scan_Vertipaq: Fact Sale'[Quantity]
            'Fact Sale'[Quantity]:
```

Physical Plan

The Physical Plan shown in Listing 11-3 doesn't contain any operators or instructions on filtering by any of the RLS filtering rules for the role. Like the Logical Plan in Listing 11-2, the Physical Plan is what you might expect to see regardless of RLS filters applied.

Listing 11-3. A simplified version of the physical plan captured when testing the calculated measure using the RLS role

```
AddColumns:
    SingletonTable:
    SpoolLookup: LogOp=Sum_Vertipaq
            ProjectionSpool<ProjectFusion<Copy>>:
                    Cache: IterPhyOp
```

DAX Query End

The Query End event carries the actual DAX query issued by Power BI Desktop to the database in the data model to obtain the result it needs for the visual. The DAX gets shown in Listing 11-4 and, like the Logical and Physical Plans, has no reference to filtering according to the active RLS role getting tested. An identical DAX Query End event gets produced even when RLS testing is turned off

Listing 11-4. The DAX Query End text for the calculated measure

```
EVALUATE
  ROW(
  "Sum_of_Quantity", 'Fact Sale'[Sum of Quantity]
)
```

VertiPaq scan

The final events to check are events relating to querying the Storage Engine (SE). RLS filtering rules turn up here in the SE events. Figure 11-14 shows a WHERE clause gets added to the pseudo-T-SQL query that is produced by the SE event.

The WHERE clause would disappear from this event when RLS role test is turned off.

```
SET DC_KIND="AUTO";
SELECT
SUM([Fact Sale (10)].[Quantity (69)]) AS [$Measure0]
FROM [Fact Sale (10)]
        LEFT OUTER JOIN [Dimension Date (19)] ON [Fact Sale (10)].[Invoice Date Key (63)]=
        LEFT OUTER JOIN [Dimension City (34)] ON [Fact Sale (10)].[City Key (59)]=[Dimensio
WHERE
        [Dimension City (34)].$ROWFILTER IN '0x0000000000000000000000000000000000000000000000
        COALESCE((PFDATAID( [Dimension Date (19)].[Calendar Year (100)] ) = 5));
```

Figure 11-14. *The Storage Engine query for the calculated measure getting tested*

As you can see in Figure 11-14, a WHERE exists and has two rows. The top row in the WHERE clause relates to the filter on the 'Dimension City' table. The text in this line of the WHERE clause uses the IN predicate and is very long. Scrolling to the very right-hand side of this text reveals that "[1818 total units, not all displayed]." The DAX query engine converts the RLS filter rule applied to the [State Province] into specific values needed to filter the internal rowid column in the 'Dimension City' table.

The second row in the WHERE clause relates to the 'Dimension Date' table and again uses an IN operator in the predicate to specify a filter over the intrinsic dictionary value for 2015.

The VertiPaq SE events captured also show the behavior of the scan when multiple RLS roles are simultaneously valid. The pseudo-T-SQL reveals a WHERE clause that concatenates a series of rules using an OR operator, rather than an AND operator. The VertiPaq SE Query End is a beneficial trace event to understand precisely what data the DAX query engine is requesting from the underlying database.

Query Plan summary

So, in summary, RLS is applied at the Storage layer and not by modifications made to DAX queries generated by Power BI Desktop (or web service) or by the queries that generate data in the Data view.

Dynamic Row-Level Security

The earlier examples in this chapter all use hardcoded values for column filtering. Sooner or later, you'll come across a requirement to apply RLS rules dynamically. A typical example is to take into account the username of the person viewing your report.

The Wide World Importers dataset has a 'Dimension Employee' table with 20 distinct values for [Employee ID]. Assume you have a requirement to lock a report, so it only shows data belonging to the user who has logged in. One way to solve this is to create 20 RLS roles with each having a hardcoded rule such as [WWI Employee ID] = 1 and then assign the appropriate individual to that role once the dataset gets published to the Power BI web service.

This approach is time consuming, has a degree of risk in that errors get made when configuring the web service, and means roles need to get managed as employees come and go.

Fortunately, a better way to solve this is by implementing dynamic RLS. There are a couple of useful DAX functions that help detect the current user and therefore implement dynamic RLS. These functions are

USERNAME

USERPRINCIPALNAME

These functions return text that carries the login details for the person currently viewing a report. The DAX expression used in an RLS filter can use these functions to compare with a column in a pre-populated table of known values of UserNames (or UserPrincipalNames). If the test finds a match, the table is filtered, along with any downstream related tables to only show rows appropriate for that username.

One approach is to pre-populate a brand-new table or add a column to an existing table with values that potentially match the output of these functions. So you need to first see what output gets generated by each function. The easiest way to test this output is to add the calculated measure in Listing 11-5 to a data model and then in a report visual that displays text.

Listing 11-5. A calculated measure to show the output of the USERNAME and USERPRINCIPALNAME name functions

```
Dynamic RLS test =
    COMBINEVALUES(
            " , " ,
            "USERNAME() = " & USERNAME() ,
            "USERPRINCIPALNAME() = " & USERPRINCIPALNAME()
    )
```

When the [Dynamic RLS test] calculated measure is added to the report canvas in Power BI Desktop, you probably see each function output a value using the *domain\ username* pattern.

It's important to publish the same report to the web service and check the value shown when viewed after logging into the web service. The value should now be different and more likely to be a pattern such as username@yourcompanydomain.com. This example is the pattern you need to use when prepopulating tables in your data model to be filtered using RLS filters and not the *domain\username* pattern.

The next step is to add these usernames to your data model. To fill the requirement using the Wide World Importers dataset, add the Calculated Column in Listing 11-6 to the 'Dimension Employee' table.

Listing 11-6. A Calculated Column to add a value to be used as a filter for dynamic RLS

```
UserPrincipalName =
    SUBSTITUTE('Dimension Employee'[Employee]," ",".") &
    "@wideworldimporters.com"
```

The code in Listing 11-6 generates a new column that creates a value designed to match the output of the USERPRINCIPALNAME() function when in the Power BI web service. The pattern shown in this Calculated Column is not the same for every organization, so tweak as required. Figure 11-15 shows a sample table using the code at Listing 11-6.

Employee Key ▼	WWI Employee ID ▼	Employee ▼	UserPrincipalName ▼
1	14	Lily Code	Lily.Code@wideworldimporters.com
2	4	Isabella Rupp	Isabella.Rupp@wideworldimporters.com
3	11	Ethan Onslow	Ethan.Onslow@wideworldimporters.com
4	7	Amy Trefl	Amy.Trefl@wideworldimporters.com
5	19	Jai Shand	Jai.Shand@wideworldimporters.com
6	8	Anthony Grosse	Anthony.Grosse@wideworldimporters.com

Figure 11-15. *The 'Dimension Employee' table with the Calculated Column from Listing 11-6*

The next step is to create a new RLS role in the data model that applies a filter to the 'Dimension Employee' table. In Figure 11-16, a new role called "Dynamic RLS Role" and a simple DAX expression are added.

The [UserPrincipalName] text with square brackets references a column in the highlighted 'Dimension Employee' table, while UserPrincipalName() with curved parentheses to the right of the text is a reference to the DAX function.

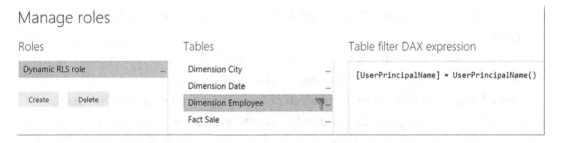

Figure 11-16. *A basic RLS role designed to filter the 'Dimension Employee' table by the USERPRINCIPALNAME function*

The new role can be tested using the View as Roles feature, and any user can be simulated using this functionality. Figure 11-17 shows the View as Roles dialog configured for testing. In this dialog, both the "Other user" and "Dynamic RLS role" are selected.

When the box next to "Other user" is checked, a text box appears allowing you to enter some text. Whatever text you enter here is reproduced by both the USERNAME and USERPRINCIPALNAME functions inside Power BI Desktop.

The ability to enter any text next to "Other user" allows you to simulate any user in Power BI Desktop mode.

Figure 11-17. *The View as Roles dialog configured for testing dynamic RLS*

In Figure 11-17, the "Other user" is set to Lily.code@wideworldimporters.com, and both the "Other user" and "Dynamic RLS role" options get ticked which means they both become active in Power BI Desktop.

The effect of having both roles active gets seen in the Report and Data views. In the Data view, the 'Dimension Employee' table now shows only the 18 rows where the new [UserPrincipalName] column matches the value entered in the View as Roles dialog.

The 'Fact Sale' table now only shows 22,642 rows, rather than 228,265 before RLS got applied. The rows visible in 'Fact Sale' get restricted to only those belonging to employee Lily Code.

All visuals on any report page now get automatically filtered to just these rows. No modifications get made to any calculated measures or columns.

The advantage of this approach is when new users get added to the 'Dimension Employee' table, nothing should need to be done to the Power BI report for the new users to access the report using filtering, aside from adding them to the appropriate App Workspace.

Testing in the Power BI web service

The report is now ready to be published to the Power BI web service. Ideally, you publish the report to an App Workspace. Once the report exists in an App Workspace, you can navigate to the Datasets area of the App Workspace and click the Security option from the context menu that appears when you click the ellipsis as per Figure 11-18.

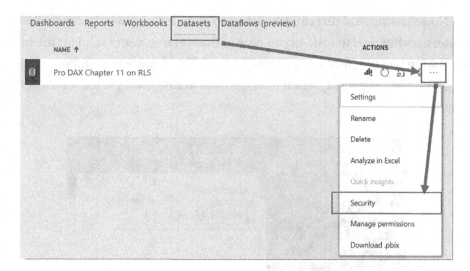

Figure 11-18. *The Datasets screen for an App Workspace in the Power BI web service*

The Security option loads the Row-Level Security screen for the selected dataset and gets shown in Figure 11-19. This screen lets you configure RLS for the dataset in the web service. Clicking the ellipsis next to the role name allows you to test the report.

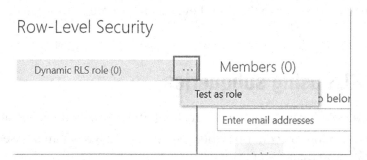

Figure 11-19. *How to invoke the Test as role feature in the Power BI web service*

Once you invoke the Test as role feature in the web service, the report appears with RLS active and set to be filtered by your credentials. This feature gets shown in Figure 11-20. A blue bar appears at the top of the report that allows you to turn off testing by clicking the "Back to Row-Level Security" arrow.

The "Now viewing as: " text can be clicked to open a dialog window that allows you to enter another user to test. As you type in the text box, users valid for the tenant appear allowing you to select the appropriate user. This feature is useful when checking what each user sees, especially when filtering rules across multiple roles start getting complicated.

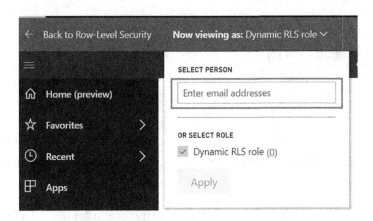

Figure 11-20.   *Testing using different users*

Once testing gets completed, you can click the "Back to Row-Level Security" text in the blue bar to navigate back to the Row-Level Security configuration window for the dataset.

Dynamic RLS using subqueries

The USERPRINCIPALNAME function is excellent for filtering data to specific users. Sometimes there are other scenarios where a dynamic RLS role can be used to minimize the overall number of roles needed.

Subqueries can get used in place of hardcoded values in RLS filters.

RLS filter can include most DAX expressions as long as they resolve to a Boolean true or false. The example shown in the filter panel in Figure 11-21 shows a filter using the MAXX function.

The IN operator can get used in conjunction with any DAX function that returns a single-column table such as SELECTCOLUMNS, TOPN, or VALUES. The VALUES function can get wrapped with a filter to provide additional flexibility.

TOPN does not calculate values in related tables, so queries that try and limit the 'Dimension City' table to only show the top ten [State Province] values based on a column in the related 'Fact Sale' table do not work.

This limitation is probably due to the implementation of RLS filtering at the Storage Engine level that cannot filter using different tables. These scenarios can get solved in data modeling by pre-computing a column in the 'Dimension City' table to show [Is Top 10 State Province By Quantity] as 1 or 0. Once this type of column gets added, it can get used quickly and effectively in an RLS filter.

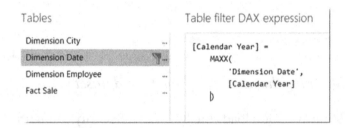

Figure 11-21. *Subquery used as part of an RLS filter*

Users added to any role with the filter suggested in Figure 11-21 are limited to only the latest year unless they also get included in other roles, in which case they also see data permitted by the other roles.

If data gets added to the 'Dimension Date' that represents a newer [Calendar Year], RLS dynamically applies the new filter state to all users assigned to the role without requiring any direct configuration changes.

Assigning users to roles

Once at least one role gets defined in the data model and the report published to the web service, users or groups can be assigned to roles and therefore become subject to RLS filtering when they view a report.

To assign users or groups to a role, find the dataset in the App Workspace, click the ellipsis, and choose the Security option, as per Figure 11-22.

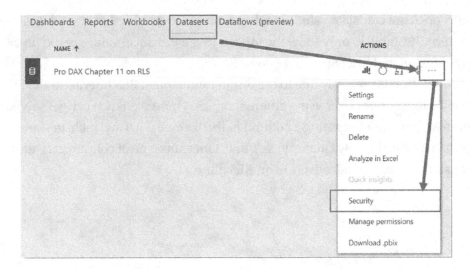

Figure 11-22. *Accessing the Security option for a dataset in an App Workspace*

Clicking the Security option loads the Row-Level Security configuration screen that displays a list of all the roles defined within the model. Filtering rules for each role cannot be modified in this view and can only get changed in Power BI Desktop.

Figure 11-23 shows the Row-Level Security screen where you can add/remove users or groups to/from each role. Once you have finished making changes, you click the save button to commit the updates.

Figure 11-23. *The add/remove users or groups screen for RLS*

Once you start typing in the text box, a valid list of users that belong to the tenant appears allowing you to select the correct users (or groups). Azure Active Directory groups can get assigned to a role which significantly reduces the amount of administration if your Power BI tenant gets connected to a fully maintained Azure Active Directory.

Note   RLS filtering gets ignored if the user is an admin of an App Workspace.
RLS only applies to read-only members.

RLS summary

Row-Level Security is a handy and often critical feature for enterprise data modeling.
It is flexible enough to cater to many scenarios. RLS can potentially speed up a report,
but equally, complex filtering rules may incur a performance hit. RLS can be applied to
an existing report and does not require actual calculations to be updated to get working.

This chapter has shown you that RLS is role based, with filtering rules for roles
defined and managed using Power BI Desktop. Once a pbix file gets published to the web
service, there is a separate process for assigning users (or groups) to roles.

Filtering rules for roles can become complicated, so this chapter has shown how RLS
can be tested both in Power BI Desktop and in the web service.

Roles can be dynamic to help reduce the need to automatically update your pbix
file each time a user starts or changes internal roles. Queries can be used to cater to
sophisticated requirements.

Users can exist in multiple roles at the same time, and in this case, the user sees all
data valid for each role they qualify to see.

PART IV

Debugging and Optimization

DAX Studio

Introduction

First of all, DAX Studio is **the** tool, maybe the only tool that helps to optimize the performance of a measure, and it even helps to improve the overall performance of a report at the whole or a specific report page.

As essential as it may be, we do not have to forget that DAX Studio is free of any cost. The people behind DAX Studio are investing immensely in DAX Studio. So for this reason, whenever something happens that makes you restart DAX Studio, maybe more than once in a performance optimization session, please keep calm and remember that without DAX Studio, you might not even start the optimization session. It's also essential to note that DAX Studio is by no means a flawed tool, but it's software made by mere mortals. This means something might happen.

DAX Studio can be downloaded from here:

`www.sqlbi.com/tools/dax-studio/`

The source code is available here:

`https://github.com/DaxStudio/DaxStudio`

The full documentation is here:

`https://daxstudio.org/`

Whenever you meet Darren Gosbell or the guys from sqlbi.com (Marco Russo and Alberto Ferrari), go and tell them how much DAX Studio has been a companion on your path to understand and master DAX. As intricate or mysterious as the inner workings of DAX may seem, don't hesitate to create a report page with just one visual to understand what's going on; maybe sometime you will discover a DAX function used by the visual that is currently not documented.

© Philip Seamark, Thomas Martens 2019
P. Seamark and T. Martens, *Pro DAX with Power BI*, https://doi.org/10.1007/978-1-4842-4897-3_12

What to expect

You may have seen a lot of videos where people are using DAX Studio to solve performance issues with reports and DAX statements. Please be aware that it is not DAX Studio that solves all these performance issues alone; it's the user that solves the problem. DAX Studio will be of tremendous help, as it helps to discover what's going on and to single out the performance bottlenecks.

The empty report page

There are many reasons to use DAX Studio:

- Optimizing the overall performance of a report page

- Optimizing the performance of a specific measure

- Searching for what's going on behind the scenes

But no matter what is the exact reason, it all starts with an empty report page. The reason behind this is that each visual that is visible on the report page creates its own DAX query. For this reason, it's necessary to start the pbix with an empty report page, because no query will be created. Connect DAX Studio to the open Power BI file, enable the All Queries trace flag, and then if DAX Studio is set up switch to the report page.

Connect to a Power BI Desktop file

As Power BI Desktop is moving fast, also the development of DAX Studio is moving fast. For this reason, it can be possible that you will discover some slight differences in screenshots, but as this is not the official documentation, we tried to make sure to use the latest version possible to take screenshots.

Nevertheless, to connect to a pbix file, it's necessary that this file is open. Before you can connect to a pbix file, this file has to be open. Make sure the file started with an empty report page. If the report did not start with an empty report page, switch to an empty one, or create one. Save the file, close Power BI Desktop, and open the file again .

Note   The Power BI file "CH12 – a simple report – large.pbix" is used.

Make sure the report started with the report page "Blank page" being active.

If DAX Studio asks for a connection, it's possible to connect all kinds of Tabular data model, but as this is about Power BI Desktop, please choose the file "CH12 – a simple report – large.pbix". This is shown in Figure 12-1.

Connect

Data Source

○ PowerPivot Model ⑦

● PBI / SSDT Model   📊 CH12 - a simple report - large   ⌄

○ Tabular Server   powerbi://api.powerbi.com/v1.0/myorg/v2 workspa ⌄

⌄ Advanced Options

Connect   Cancel

Figure 12-1. *Connect to pbix*

Figure 12-2 will explain in short what you will see after a successful connection.

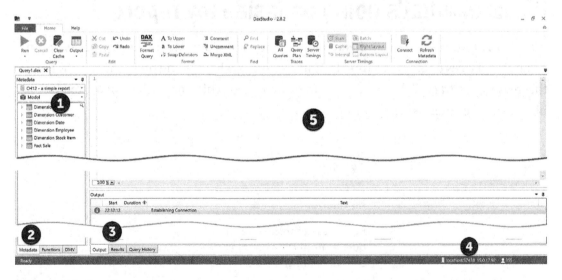

Figure 12-2. *First contact*

I'm Here	A Short Explanation
①	The data model.
②	Available views: Metadata – Provides insight into the data model. Functions – List of available DAX functions that can be used. DMV – List of Dynamic management views. Please be aware that not all DMVs can be used in combination with a data model inside a Power BI Desktop file.
③	Tabs: This table will provide different perspectives on a DAX query.
④	The server name of the Analysis Services engine instance that runs inside Power BI Desktop. This server name can also be used to connect the SQL Server Management Studio to this database. Using SQL Server Profiler, it can also be used to trace what's happening inside a Power BI report.
⑤	The currently empty section that allows to write/change DAX statements. This is the editor window.

Discover what's going on inside my report

After DAX Studio has been connected to the Power BI report (to be more precise, to the Analysis Services engine), it's time to start exploring what's going on in the report. For this reason, the trace "All Queries" has to be activated. This trace will be activated by hitting the "All Queries" icon in the menu bar.

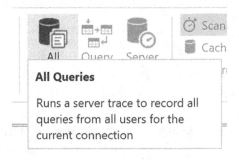

The activation of the trace is visualized, as the icon is now highlighted. The activation of this trace type is also shown in the Output tab (Figure 12-3).

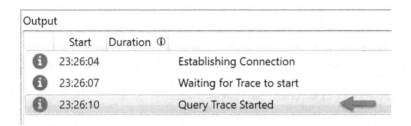

Figure 12-3. *All Queries trace activated*

Simple visuals

To capture the queries that are executed when the report page will be activated, activate
the report page "Simple Visuals."

After activating the report page, the DAX queries will show up in DAX Studio
(Figure 12-4).

StartTime	Type	Duration	User	Database	Query
11:44:47	DAX	9	tmart	CH12 - a...	EVALUATE ROW("SumTax_Amount", CALCULATE(SUM('Fact Sale'[Tax Amount])))
11:44:47	DAX	18	tmart	CH12 - a...	EVALUATE ROW("SumQuantity", CALCULATE(SUM('Fact Sale'[Quantity])), "SumProfit", CALCULATE(SU
11:44:47	DAX	8	tmart	CH12 - a...	EVALUATE ROW("SumProfit", CALCULATE(SUM('Fact Sale'[Profit])))
11:44:47	DAX	6	tmart	CH12 - a...	EVALUATE ROW("SumTax_Rate", CALCULATE(SUM('Fact Sale'[Tax Rate])))
11:44:47	DAX	13	tmart	CH12 - a...	EVALUATE ROW("SumProfit", CALCULATE(SUM('Fact Sale'[Profit])), "SumQuantity", CALCULATE(SUM('
11:44:47	DAX	27	tmart	CH12 - a...	EVALUATE ROW("SumQuantity", CALCULATE(SUM('Fact Sale'[Quantity])))

Output | Results | Query History | ▶ All Queries

Figure 12-4. *Each visual has its own query*

Using the "Copy All" icon copies all the trace results into the editor window
(Figure 12-5).

```
 1
 2 // DAX query against Database: CH12 - a simple report
 3 EVALUATE
 4   ROW(
 5     "SumTax_Amount", CALCULATE(SUM('Fact Sale'[Tax Amount]))
 6   )
 7
 8 // DAX query against Database: CH12 - a simple report
 9 EVALUATE
10   ROW(
11     "SumQuantity", CALCULATE(SUM('Fact Sale'[Quantity])),
12     "SumProfit", CALCULATE(SUM('Fact Sale'[Profit])),
13     "SumTax_Amount", CALCULATE(SUM('Fact Sale'[Tax Amount])),
14     "SumTax_Rate", CALCULATE(SUM('Fact Sale'[Tax Rate]))
15   )
16
17 // DAX query against Database: CH12 - a simple report
18 EVALUATE
19   ROW(
20     "SumProfit", CALCULATE(SUM('Fact Sale'[Profit]))
21   )
22
23 // DAX query against Database: CH12 - a simple report
24 EVALUATE
25   ROW(
26     "SumTax_Rate", CALCULATE(SUM('Fact Sale'[Tax Rate]))
27   )
28
29 // DAX query against Database: CH12 - a simple report
30 EVALUATE
31   ROW(
32     "SumProfit", CALCULATE(SUM('Fact Sale'[Profit])),
33     "SumQuantity", CALCULATE(SUM('Fact Sale'[Quantity])),
34     "SumTax_Amount", CALCULATE(SUM('Fact Sale'[Tax Amount])),
35     "SumTax_Rate", CALCULATE(SUM('Fact Sale'[Tax Rate]))
36   )
37
38 // DAX query against Database: CH12 - a simple report
39 EVALUATE
40   ROW(
41     "SumQuantity", CALCULATE(SUM('Fact Sale'[Quantity]))
42   )
43
44
```

Figure 12-5. *Trace result simple visuals*

At the moment, the key takeaways are

- Each visual is populated by its query.

- Each query returns a table, even the queries that are populating a card visual.

A Power BI report page and the filter pane

In November 2018, the product team introduced the new filter experience to Power BI Desktop. Next to regaining valuable screen estate by placing slicers, another essential capability of the filter pane slipped in almost unnoticed. As already mentioned, each "visible" visual creates its own DAX query. The columns that are added to the new filter pane will not be evaluated when the report is started.

Listing 12-1 shows the DAX queries that are generated if the report page "Simple Visuals – slicer" is activated.

Listing 12-1. DAX queries – report page "Simple Visuals – slicer.dax"

```
// DAX query against Database: CH12 - a simple report
EVALUATE
  ROW(
    "SumProfit", CALCULATE(SUM('Fact Sale'[Profit])),
    "SumQuantity", CALCULATE(SUM('Fact Sale'[Quantity])),
    "SumTax_Amount", CALCULATE(SUM('Fact Sale'[Tax Amount])),
    "SumTax_Rate", CALCULATE(SUM('Fact Sale'[Tax Rate]))
  )

// DAX query against Database: CH12 - a simple report
EVALUATE
  TOPN(
    101,
    VALUES('Dimension Customer'[Customer Key]),
    'Dimension Customer'[Customer Key],
    1
  )

ORDER BY
  'Dimension Customer'[Customer Key]

// DAX query against Database: CH12 - a simple report
EVALUATE
  ROW(
    "SumQuantity", CALCULATE(SUM('Fact Sale'[Quantity]))
  )
```

```
// DAX query against Database: CH12 - a simple report
EVALUATE
  ROW(
    "SumTax_Rate", CALCULATE(SUM('Fact Sale'[Tax Rate]))
  )

// DAX query against Database: CH12 - a simple report
EVALUATE
  ROW(
    "SumTax_Amount", CALCULATE(SUM('Fact Sale'[Tax Amount]))
  )

// DAX query against Database: CH12 - a simple report
EVALUATE
  ROW(
    "SumProfit", CALCULATE(SUM('Fact Sale'[Profit]))
  )

// DAX query against Database: CH12 - a simple report
EVALUATE
  ROW(
    "SumQuantity", CALCULATE(SUM('Fact Sale'[Quantity])),
    "SumProfit", CALCULATE(SUM('Fact Sale'[Profit])),
    "SumTax_Amount", CALCULATE(SUM('Fact Sale'[Tax Amount])),
    "SumTax_Rate", CALCULATE(SUM('Fact Sale'[Tax Rate]))
  )
```

Listing 12-1 contains a DAX query that populates the slicer on the report page. This DAX query is **highlighted** in this listing.

If the DAX queries are captured from the report page "Simple Visuals – Filter Pane," it's essential to note that no DAX statement is executed to populate the slicer or to be precise the filter added to the page-level section of the filter pane.

Figure 12-6 shows that a value is selected from the filter.

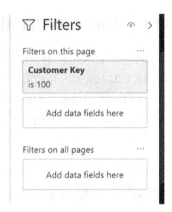

Figure 12-6. *Filter pane with selected value*

If the queries are captured from the report page, with a value selected as the Customer Key is 100, it's essential to note that just six queries will be created/executed. Listing 12-2 shows only one of these six queries.

Listing 12-2. DAX queries – report page "Simple Visuals – filter pane selected value.dax"

```
DEFINE VAR __DSOFilterTable =
  TREATAS({"100"}, 'Dimension Customer'[Customer Key])

EVALUATE
  SUMMARIZECOLUMNS(
    __DSOFilterTable,
    "SumTax_Amount", IGNORE(CALCULATE(SUM('Fact Sale'[Tax Amount])))
  )
```

The important thing to notice is the following: Instead of executing a query to populate the slicer, just a tiny little string is added to each query that defines a filter table and is used in the query to retrieve the data for the visual. The lines that define the filter table are highlighted (bold).

Tip   Move the slicer to the filter pane that is not as often used as you might think! This can improve the overall performance of the report/report page tremendously.

Report- and page-level filter: A word of warning

Quite often, the data model contains some data that should not be considered on a report page or at the report at all. For this reason, it's possible to exclude specific values from a column by defining page-level filter or report-level filter. Figure 12-7 shows that two Sales Territories are excluded; this figure refers to the report page "Report and Page Level Filter."

Caution   All the previous screenshots and listings did not consider a report-level filter; for this reason, the report filter has to be added to follow this section.

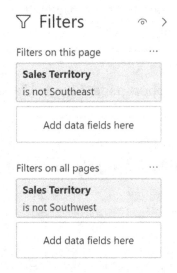

Figure 12-7. *Report- and page-level filter*

Listing 12-3 shows all the queries that have been created by activating the report page "Report and Page Level Filter." This listing contains all the four queries that are executed to populate the visuals on the report page.

Listing 12-3. DAX queries – "report and page level filter.dax"

```
// DAX query against Database: CH12 - a simple report
DEFINE VAR __DS0FilterTable =
  FILTER(
    KEEPFILTERS(VALUES('Dimension City'[Sales Territory])),
```

```
    AND(
      'Dimension City'[Sales Territory] <> "Southeast",
      'Dimension City'[Sales Territory] <> "Southwest"
    )
  )

EVALUATE
  SUMMARIZECOLUMNS(
    __DSOFilterTable,
    "Quantity___All_Sales_Territories", IGNORE('Fact Sale'[Quantity - All
    Sales Territories])
  )

// DAX query against Database: CH12 - a simple report
DEFINE VAR __DSOFilterTable =
  FILTER(
    KEEPFILTERS(VALUES('Dimension City'[Sales Territory])),
    AND(
      'Dimension City'[Sales Territory] <> "Southeast",
      'Dimension City'[Sales Territory] <> "Southwest"
    )
  )

EVALUATE
  SUMMARIZECOLUMNS(
    __DSOFilterTable,
    "Quantity___Values_Sales_Territories_", IGNORE('Fact Sale'[Quantity -
    Values(Sales Territories)])
  )

// DAX query against Database: CH12 - a simple report
DEFINE VAR __DSOFilterTable =
  FILTER(
    KEEPFILTERS(VALUES('Dimension City'[Sales Territory])),
```

```
    AND(
      'Dimension City'[Sales Territory] <> "Southeast",
      'Dimension City'[Sales Territory] <> "Southwest"
    )
  )

EVALUATE
  TOPN(
    502,
    SUMMARIZECOLUMNS(
      ROLLUPADDISSUBTOTAL('Dimension City'[Sales Territory],
      "IsGrandTotalRowTotal"),
      __DS0FilterTable,
      "Total_Quantity", 'Fact Sale'[Total Quantity],
      "quantity_divided_by_all", 'Fact Sale'[quantity divided by all],
      "quantity_divided_by_allselected", 'Fact Sale'[quantity divided by
      allselected],
      "quantity_divided_by_values", 'Fact Sale'[quantity divided by values]
    ),
    [IsGrandTotalRowTotal],
    0,
    'Dimension City'[Sales Territory],
    1
  )

ORDER BY
  [IsGrandTotalRowTotal] DESC, 'Dimension City'[Sales Territory]

// DAX query against Database: CH12 - a simple report
DEFINE VAR __DS0FilterTable =
  FILTER(
    KEEPFILTERS(VALUES('Dimension City'[Sales Territory])),
    AND(
      'Dimension City'[Sales Territory] <> "Southeast",
      'Dimension City'[Sales Territory] <> "Southwest"
    )
  )
```

```
EVALUATE
  SUMMARIZECOLUMNS(
    __DSOFilterTable,
    "Quantity__ALLSELECTED_Sales_Territory_", IGNORE('Fact
    Sale'[Quantity - ALLSELECTED(Sales Territory)])
  )
```

Both filters, the page-level filter Sales Territory is not Southeast and the report-level filter Sales Territory is not Southwest, are combined to a single filter table __DSOFilterTable. This combination is done by these lines:

```
DEFINE VAR __DSOFilterTable =
  FILTER(
    KEEPFILTERS(VALUES('Dimension City'[Sales Territory])),
    AND(
      'Dimension City'[Sales Territory] <> "Southeast",
      'Dimension City'[Sales Territory] <> "Southwest"
    )
  )
```

This must be noticed: DEFINE has to be used before a variable is defined; the definition section ends with the next EVALUATE statement. The filter table contains all the values from the column 'Dimension City'[Sales Territory] in the current filter context, except the values excluded by the page- and report-level filter. This filter table is used in each query.

Warning   Whenever the ALL function is used, the values excluded by the page- or report-level filter will be included again. For this reason, whenever a report-level filter is used to either exclude or to specify one or more values, take a moment and consider to remove these elements from the dataset during the data load. Report- and page-level filter in combination with the ALL function can become error-prone. To rephrase this, ALL does not consider the report- or page-level filter.

Time Intelligence: Auto date/time

Whenever DAX Studio is connected to a pbix file, DAX Studio shows not just the obvious tables, but also some system tables. These system tables are highlighted in Figure 12-8.

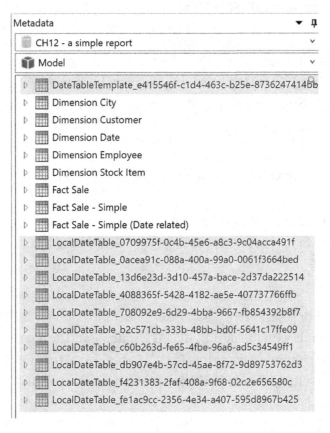

Figure 12-8. *Date columns and date tables*

At the moment, Power BI creates a dedicated date table for each column of the data type date or datetime. If there are two date columns in one table, two tables will be created. If there are another two columns of these data types in a different table, another two date tables will be created. This functionality is considered a bad "habit." This is because of the following

- Memory consumption

- Load time of the data model

The memory consumption can be neglected as the data model inside Power BI desktop just consumes ~834 MB of RAM.

The time to load and refresh a model is much more important, especially when there are a lot of date/datetime columns inside the data model. Even if these system-generated tables do not do any harm, it's a well-known best practice to not rely on these system date tables.

It seems that these tables have an advantage in comparison to a dedicated calendar table as they allow to use a date hierarchy whenever a date column is used on a visual. This hierarchy is defined in the DateTableTemplate, but hierarchies can be defined as well in a dedicated calendar table.

To disable the creation of these system date tables, it's necessary to disable the setting "Auto date/time." This is shown in Figure 12-9.

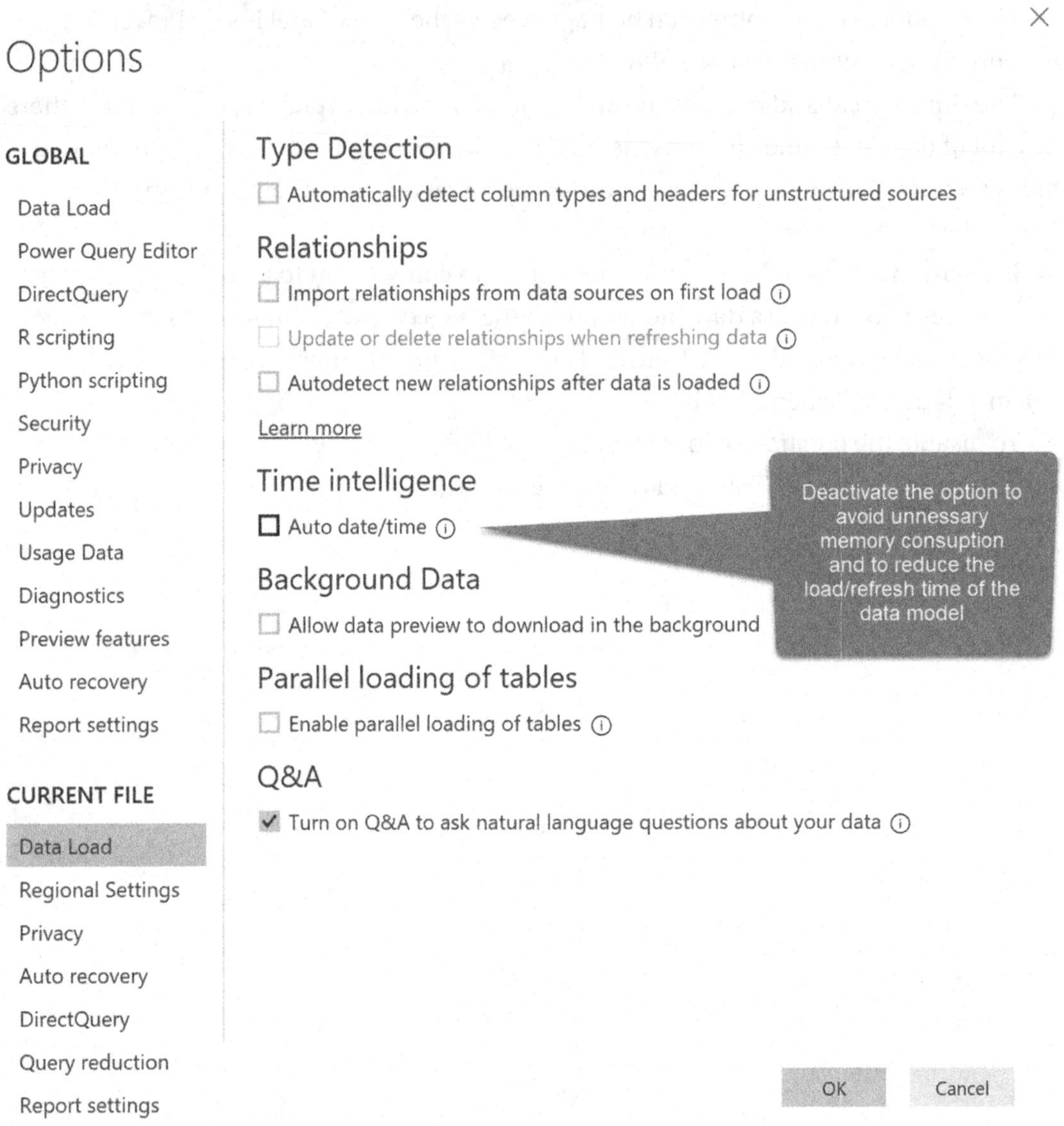

Options

GLOBAL

Data Load

Power Query Editor

DirectQuery

R scripting

Python scripting

Security

Privacy

Updates

Usage Data

Diagnostics

Preview features

Auto recovery

Report settings

CURRENT FILE

Data Load

Regional Settings

Privacy

Auto recovery

DirectQuery

Query reduction

Report settings

Type Detection

☐ Automatically detect column types and headers for unstructured sources

Relationships

☐ Import relationships from data sources on first load ⓘ

☐ Update or delete relationships when refreshing data ⓘ

☐ Autodetect new relationships after data is loaded ⓘ

Learn more

Time intelligence

☐ Auto date/time ⓘ

> Deactivate the option to avoid unnessary memory consuption and to reduce the load/refresh time of the data model

Background Data

☐ Allow data preview to download in the background

Parallel loading of tables

☐ Enable parallel loading of tables ⓘ

Q&A

☑ Turn on Q&A to ask natural language questions about your data ⓘ

OK   Cancel

Figure 12-9. *Time intelligence – auto date/time*

The DAX query editor in DAX Studio

Caution   From now on, "CH12 – a simple report.pbix" is used.

Before it's possible to start optimizing DAX statements, it's necessary to get familiar with the editor. For this reason, open the file "CH12 – a simple report.pbix" and make sure it started with the report page "Blank page"; if this is not the case, select this page, save the file, and open it again.

After the file has been opened with the "Blank page" active, start DAX Studio, connect to the Power BI Desktop file, and activate the All Queries trace.

On the left, in the Metadata tab, mark the measure Quantity – All Sales Territories and open the context menu. This is shown in Figure 12-10.

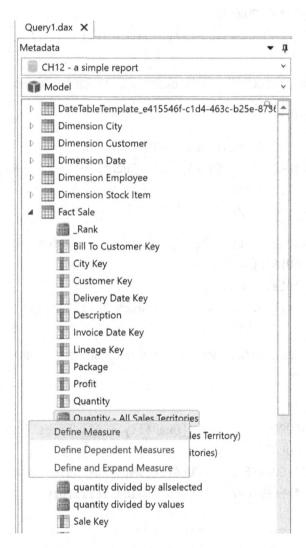

Figure 12-10. *DAX Studio editor – measure context menu*

The three commands work as follows:

Command	Action

Define Measure	Copies the definition of the measure to the editor – the result:
	
	```DEFINE```
	
	```MEASURE 'Fact Sale'[Quantity - All Sales Territories] =```
	```CALCULATE(```
	``` [Total Quantity]```
	``` ,ALL('Dimension City'[Sales Territory])```
	```)```

Define Dependent Measures	Copies the definition of the measure to the editor – the result:
	
	```DEFINE```
	
	```---- MODEL MEASURES BEGIN ----```
	```MEASURE 'Fact Sale'[Total Quantity] = CALCULATE(SUM('Fact Sale'[Quantity]))```
	```---- MODEL MEASURES END ----```
	
	```MEASURE 'Fact Sale'[Quantity - All Sales Territories] =```
	```CALCULATE(```
	``` [Total Quantity]```
	``` ,ALL('Dimension City'[Sales Territory])```
	```)```

Define and Expand Measure	Creates a single measure where all the dependent measures expanded – the result:
	
	```DEFINE```
	```MEASURE 'Fact Sale'[Quantity - All Sales Territories] =```
	```CALCULATE(```
	``` CALCULATE (CALCULATE(SUM('Fact Sale'[Quantity])))```
	``` ,ALL('Dimension City'[Sales Territory])```
	```)```

Using the measure definition

From the Metadata tab, choose "Define Dependent Measures" from the context menu for the measure quantity divide by all from the table Fact Sale. This is shown in Figure 12-11.

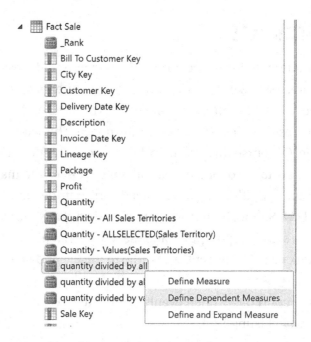

Figure 12-11. *DAX Studio editor – define dependent measures*

Figure 12-12 shows the result of this action in the query editor.

```
1  DEFINE
2
3  ---- MODEL MEASURES BEGIN ----
4  MEASURE 'Fact Sale'[Quantity - All Sales Territories] = CALCULATE(
5      [Total Quantity]
6      ,ALL('Dimension City'[Sales Territory])
7  )
8  MEASURE 'Fact Sale'[Total Quantity] = CALCULATE(SUM('Fact Sale'[Quantity]) )
9  ---- MODEL MEASURES END ----
10
11 MEASURE 'Fact Sale'[quantity divided by all] = DIVIDE([Total Quantity], [Quantity - All Sales Territories], BLANK())
12
```

Figure 12-12. *DAX Studio editor – dependent measures*

This is essential:

Note   DEFINE introduces the definition phase of measures. The keyword MEASURE starts the measure definition.

If you are wondering why a table name is used as a prefix, this corresponds to the table that the measure is assigned to. Whenever you are referencing a measure in Power BI, don't use the table name. This will confuse the reader of the DAX statement.

Query performance

The optimization of query performance most often targets to reduce the time it takes to calculate a measure. As it takes time to create DAX statements that are returning the expected and correct result, it even takes more time to create DAX statements that are returning the correct result with blazing-fast performance. This cannot be achieved by reading a book alone. Of course this may help, but without practice and more practice, not to mention the importance of practice, it will not be possible to master DAX.

To better understand what's going on, it's necessary to understand how to trace performance using DAX Studio; it's necessary to capture a query. For this reason, open the pbix file "CH12 – a simple report.pbix" and make sure that the file opens up with the report page "Blank page." Connect DAX Studio to the file and enable the All Queries trace.

Then activate the report page "open events – unrelated." We are interested in the DAX statement shown in Listing 12-4.

Listing 12-4. DAX queries – "event not so fast.dax"

```
EVALUATE
  TOPN(
    502,
    SUMMARIZECOLUMNS(
      ROLLUPADDISSUBTOTAL('Dimension Date unrelates'[Calendar Month Label],
        "IsGrandTotalRowTotal"),
      "events_in_progress___not_so_fast", 'Fact Sale'[events in progress -
      not so fast]
    ),
    [IsGrandTotalRowTotal],
    0,
    'Dimension Date unrelates'[Calendar Month Label],
    1
  )
```

```
ORDER BY
  [IsGrandTotalRowTotal] DESC, 'Dimension Date unrelates'[Calendar Month
  Label]
```

For now it's not that important to fully understand what's going on. But by selecting the "large" query (duration) by double-clicking the query, DAX Studio will look as in Figure 12-13.

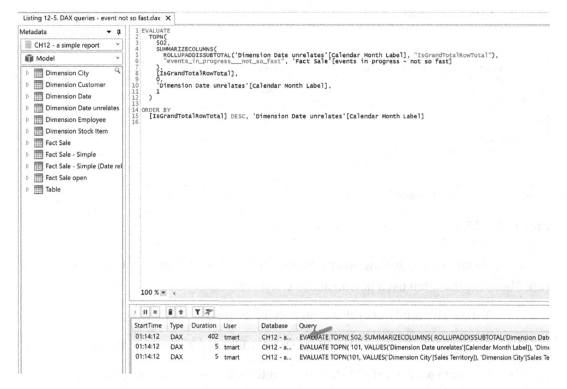

Figure 12-13. *Events – preparing server timings*

After the query of interest is captured to the query editor, it's possible to switch traces. The first trace will be Server Timings (see Figure 12-14). To activate this trace, it's necessary to deactivate the trace "All Queries" first.

Figure 12-14. *Events – server timings*

After the Server Timings trace is activated, the Server Timings output window will become available (Figure 12-15).

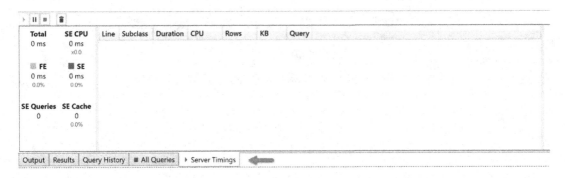

Figure 12-15. *Events – server timings output*

It's necessary to understand the performance counters that are measured for each run. These are the often used performance counters:

- Total – The total query duration in milliseconds

- FE (Formula Engine) – Duration spent utilizing the Formula Engine

- SE (Storage Engine) – Duration spent utilizing the Storage Engine

It's necessary to remember that the Storage Engine is the fast engine. This means the more time is spent inside the Formula Engine, the more query optimizing (meaning DAX optimization) becomes important.

A more detailed description can be found here: `https://daxstudio.org/documentation/features/server-timings-trace/`.

CHAPTER 13

Query Plans

Introduction to Query Plans

A handy feature of the DAX engine is the ability to access a useful piece of information called a Query Plan. A Query Plan provides insight into the inner workings of how the DAX engine goes about its task to deliver results requested by any given DAX query.

A real-world comparison might be to provide a breakdown of tasks required to complete a particular project. If the project was to build a birdhouse, a project plan could get generated before starting, being updated along the way once specific tests have been executed.

A pseudo-Query Plan for building a birdhouse might look like the following:

```
Assemble the parts using the tools of the design.
    Find a blueprint for the ideal birdhouse.
        Search the Internet for ideas.
    Obtain the parts required.
        Drive to the hardware store and collect the material.
            Pay for the material.
    Retrieve tools required from the workshop.
        Get a hammer.
        Get a saw.
        Get nails.
        Get a tape measure.
```

With this example, each line in the plan represents a subtask for the overall project. The final step for the project is listed in the top row and has the least degree of indentation. The order of tasks is not necessarily top to bottom. There are three sibling nodes specific to the tasks of finding a blueprint and getting parts and tools required. The three sibling nodes all have a series of subtasks that relate to that specific step in the lines immediately underneath, and the indentation helps to highlight the order and relationship of each subtask within the node.

© Philip Seamark, Thomas Martens 2019
P. Seamark and T. Martens, *Pro DAX with Power BI*, https://doi.org/10.1007/978-1-4842-4897-3_13

A Query Plan is particularly useful when trying to debug or optimize a slow-performing query. If you have worked with T-SQL, you may be familiar with the concept as similar Query Plans are also available when using the SQL Server database engine. When you execute T-SQL queries, you have the option to return a Query Plan along with the results of the query. The T-SQL, the Query Plan, can be viewed in the SSMS tool as a graphical depiction of the approach taken to return the results.

The T-SQL version of the Query Plan provides a breakdown of each sub-step and includes helpful information such as the time taken for each step if indices get used and the number of rows used as an input or output for each sub-step, among others.

The DAX version of the Query Plan is text-based and formatted using multiple lines and indentation to provide a basic tree-based structure. Each line in a plan represents an operation as part of the overall query. Query Plans are available to any client tool that can start and run a trace against an Analysis Services data model. Indentation and equivalent tab spacing are used to highlight sibling nodes vs. subtasks.

More on Query Plans

In DAX, three plans get generated for every query. However, only two plans get exposed as trace events. The three plans are

> Query Expression Tree
>
> Logical Plan
>
> Physical Plan

The Query Expression Tree is a straight tree map of the DAX expression and does not appear as a profiler event in a database trace. The Query Expression Tree gets converted and simplified into a Logical Plan Tree. The DAX engine may partially execute the logical tree before building a Physical Plan Tree. It is the Physical Plan that is executed to return the final dataset from the database.

When running a trace against a database, the *Query Plan* event always returns two query plans. One is the Logical Plan, while the other is the Physical Plan. The Logical Plan can help you understand the primitive operations used by higher-level functions. Scenarios, where functions are syntactic sugar shortcuts for other operations, can be identified using the Logical Plan.

The Physical Plan can help highlight areas of a query where the Formula Engine is causing performance issues.

How to find

Two free tools that can be used to access a Query Plan are DAX Studio and SQL Server Management Studio (SSMS). Both tools can run a Profiler trace against an instance of an SQL Server Analysis Services (SSAS) database. Power BI Desktop is an instance of an SSAS database.

A Profiler trace captures events that take place inside the database engine. Events come in all shapes and sizes and are not always related to a user-generated query. When starting a trace against a data model, only certain events of interest need to be selected; otherwise, the trace produces a vast amount of information making it difficult to zero in on the useful detail.

Once a trace has been started, it can be paused or stopped. The events captured can be saved as a TRC file for later review.

The first step to run a trace using SSMS is to connect to the data model of interest. The port number of the data model is required when connecting as per Figure 13-1. There are several ways to determine the local port number of an instance of Power BI Desktop. One of the easiest ways to find the port number is using DAX Studio to connect to the data model you wish to run a trace over. The port number shown in Figure 13-1 is *localhost:50086,* and this information needs to be added in full when connecting a trace to a data model.

Figure 13-1. *Shows the port number of an instance of Power BI Desktop*

The Profiler tool can be started from within SSMS from the Tools menu by clicking the SQL Server Profiler option. The profiler tool can also be started directly from the operating system menu by searching for *SQL Server Profiler* in the search area. Once the Profiler tool has started, a dialog box similar to that in Figure 13-2 appears that asks for information on how to connect to the data model.

For Power BI Desktop on a local machine, the Server type dropdown box should always be *Analysis Services*. The Server name text box needs to include the text *localhost:* along with the current port number of the instance you wish to connect. This port number changes each time Power BI Desktop gets restarted.

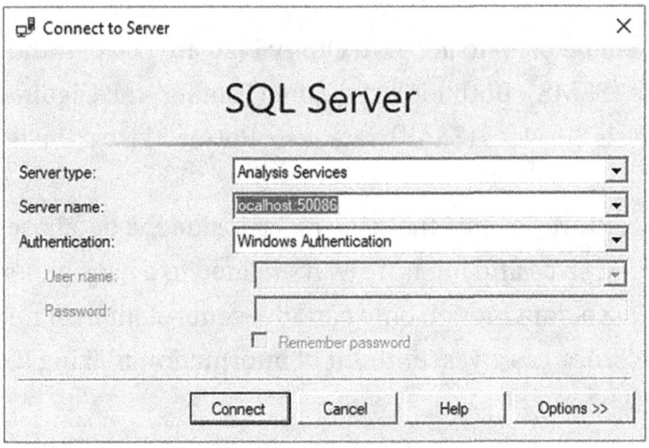

Figure 13-2. *Shows the connection dialog for the Profiler tool*

Note   Once XMLA endpoints have been opened up in the Power BI web service, you can connect the Profiler tool directly to published data models in the cloud.

Once the Profiler tool has successfully connected to a data model, a new trace can be started from the File menu using the *new trace* menu item. Clicking the *new trace* option opens a Trace Properties window. The Events Selection tab of the Trace Properties window provides a list of events that get reported while the trace is running. The recommended events to select for basic tracing are the following:

Query Events

Query End

Query Processing

DAX Query Plan

VertiPaq SE Query Cache Match

VertiPaq SE Query End

A whitepaper titled "Understanding DAX Query Plans" written in 2012 by Alberto Ferrari calls these events out as ideal events to capture when tracing Analysis Services databases. I must agree that these are an excellent starting point and additional event types should only get added in more advanced scenarios. These two query events are only for imported tables. There are different events for DirectQuery tables.

Figure 13-3 shows the Trace Properties window when configured using the recommended events.

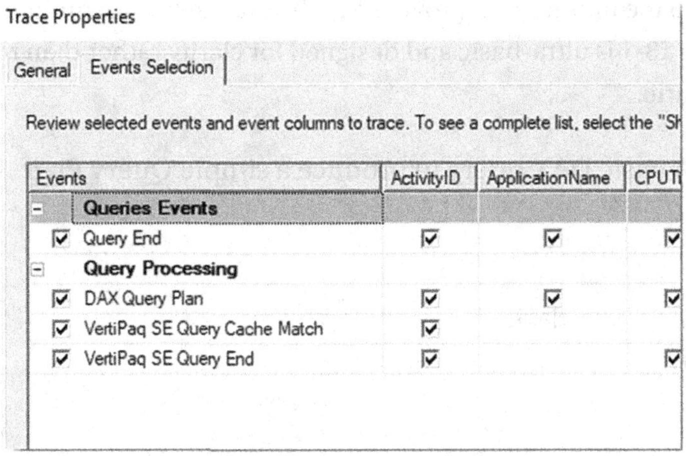

Figure 13-3. *SSMS profiler recommended Trace Properties events*

Once the events are defined, a trace can get started by clicking the Run button in the Trace Properties window. From here, any query issued to the data model adds multiple trace events to the window showing trace activity. The trace window includes user activity in the report canvas area of Power BI Desktop as well as DAX queries issued directly against the data model using SSMS or DAX Studio.

It's possible to rearrange the columns in the trace window, so useful columns get grouped toward the left-hand side of the window. The following are handy columns to move to the left of the column order:

EventClass

EventSubclass

Duration

CPUTime

Note   Trace events can be saved as a trace template to allow for quick retrieval of the set of events for future debugging.

The basic plan (my first plan)

A simple DAX query is used to introduce the structure and basic layout of a DAX query plan to help keep the information provided by the plan to a bare minimum. The query shown in Listing 13-1 is ultra-basic and designed for clarity rather than reflecting a more real-world scenario.

Listing 13-1. A basic DAX query to produce a simple Query Plan

```
EVALUATE
    ROW(
            "My Column Name",
            1 + 1
            )
```

Figure 13-4 shows both the query in SSMS and the expected result of "2" using the query from Listing 13-1.

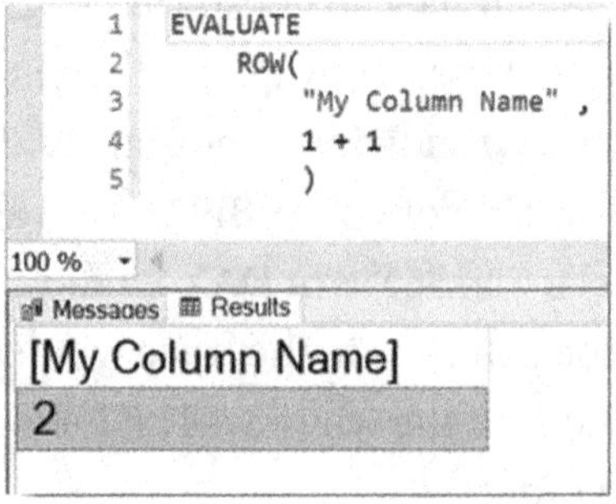

Figure 13-4. *Basic DAX query and result using SSMS*

The following three events get reported when the query from Listing 13-1 runs against a data model with an active trace. Two of the three events are DAX Query Plan, while the third is a Query End event. The detail in the EventSubClass column shows which of the two DAX Query Plan events are the Logical and Physical plans.

The full detail of each query plan is shown in the bottom section of the trace window when an individual DAX Query Plan line gets selected. The same information also gets shown in the TextData column.

The Logical Plan (my first plan)

The structure of each line in a Logical Plan follows the same pattern. The first item of text up to the first colon is the operator name. All text after the first colon represents the operator properties with the first item of the operator properties showing the operator type.

There are only two Logical Plan operator types, as shown in Table 13-1. The ScaLogOp type always returns a single value. The RelLogOp type typically deals with operations involving multiple columns that aren't always on the same table. The RelLogOp operator type can often have child steps that use the ScaLogOp operator type. The ScaLogOp operator type can have RelLogOp as children too.

Note   One of the most popular, RelLogOp is derived from Values('Table'[Column]) which is a single column. The difference between ScaLogOp and RelLogOp is that the former returns a value and the latter returns a table, regardless of the number of columns and how many physical tables are involved.

Table 13-1. *Operator types for Logical Plans*

Operator Type	Description
ScaLogOp	Scalar logical operatorReturns a single value (text, numeric, Boolean, and so on)
RelLogOp	Relational logical operatorReturns a table of values

The Logical Plan generated by the query from Listing 13-1 is shown in Listing 13-2.

Listing 13-2. The Logical Plan returned for the query in Listing 13-1

```
AddColumns: RelLogOp DependOnCols()() 0-0 RequiredCols(0)("[My Column Name])
    Constant: ScaLogOp DependOnCols()() Integer DominantValue=2
```

In the example shown in Listing 13-2, the Logical Plan consists of two lines. The first item of text in each line is the operator name. For the first line, the operator name is *AddColumns*, and for the second line, the operator name is *Constant*. These represent the underlying primitive functions used by DAX to satisfy each aspect of the query.

The operator type for the first line is RelLogOp. This operator type combined with the operator name of AddColumns suggests these belong to the ROW() function in the query in Listing 13-1 and relate to the work to return the final table.

The operator type of ScaLogOp on the second line, combined with the operator name of Constant, suggests this is the line that relates to the simple math operation.

The query doesn't reference any table or column from the data model, so no events related to the Storage Engine get reported as part of the trace.

The additional text in the operator properties provides extra detail that is relevant to each operator.

The Physical Plan (my first plan)

The structure for the Physical Plan is the same as the Logical Plan. The first text in each line represents the operator name. There are many operators in a Physical Plan, and these provide a clue as to which internal DAX function gets used to satisfy each aspect of a query. Table 13-2 shows the main operator types for a Physical Plan.

The text after the first semicolon relates to the operator properties for the line, with the first item after the semicolon declaring the operator type for the operator. The Physical plan can only have two types of operators.

Table 13-2. *Operator types for Physical Plans*

Operator Type	Description
LookupPhyOp	Lookup physical operatorReturns a single scalar value taking the current row as input
IterPhyOp	Iterator physical operatorReturns a sequence of rowsCan use the current row as an optional input

The Physical Plan generated by the query in Listing 13-1 is shown in Listing 13-3.

Listing 13-3. The Physical Plan returned for the query in Listing 13-1

```
AddColumns: IterPhyOp LogOp=AddColumns IterCols(0)("[My Column Name])
    SingletonTable: IterPhyOp LogOp=AddColumns
    Constant: LookupPhyOp LogOp=Constant Integer 2
```

This Physical Plan has three lines. The top line represents the final output for the query. There are two subnodes for the query that exist at the same level. The SingletonTable operator has a type of IterPhyOp, which means it potentially returns a series of rows. The third and bottom line of the plan tells us the operator is Constant and has a type of LookupPhyOp which means it returns a single value. The additional properties for this line specify the value is 2 (the result of 1+1) and that the data type for the Constant is Integer.

Query times

> **Note**   The DAX Query Plan event type does not report a value for the duration in the profiler trace.

Neither the Logical nor the Physical Query Plan events report a figure for the duration, which means these events alone are not always useful for debugging long-running queries. Often the combination of both DAX Query Plans *and* other events helps isolate and identify the reason behind a poor-performing query.

Clear cache

A critical tip when debugging DAX queries for performance is the need to clear the DAX engine cache between queries so that a fair comparison can get made. Running the same query in quick succession without clearing the cache can skew duration values due to subsequent executions taking advantage of cached results.

The xVelocity engine keeps a small cache which can be observed by including the *VertiPaq SE Query Cache Match* event in a Profiler trace. Ideally, you do not see this event occur when doing performance testing. This event indicates the query has taken advantage of a previous execution of the same query.

Clearing the cache requires passing an XMLA query to the data model. Running the XMLA query can be performed in either SSMS or DAX Studio.

Using SSMS

To clear the data model cache using SSMS, start a new XMLA query type and connect to the data model in question. A new XMLA query type session starts by clicking the XMLA button in the New Query menu item as per Figure 13-5.

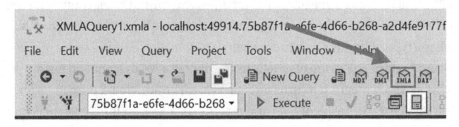

Figure 13-5. *Using the SSMS menu to start an XMLA connection to a data model*

XMLA commands can get applied from any of the query types (MDX, DMX, XMLA, and DAX); however, when used in a DAX or MDX connection window, GO statements can be inserted between ClearCache blocks and DAX EVALUATE statements to batch and run multiple DAX statements at the same time. An example of code using the GO statement in SSMS is shown in Figure 13-8.

Separating multiple DAX queries with GO and ClearCache statements is an excellent way to debug and diagnose similar queries by executing them in the same batch. This technique also helps prevent accidentally forgetting to clear the cache and getting distracted by incorrect results regarding performance. Figure 13-8 shows an example of a ClearCache command used in conjunction with a DAX query. Be sure to add an XMLA ClearCache statement before every DAX statement in a batch.

Once the XMLA connection gets successfully made, the ClearCache query from Figure 13-6 can get issued. Note the value inside the DatabaseID element needs to be updated to reflect each data model. Do not use the DatabaseID shown in Figure 13-6. If the data model is a Power BI Desktop file, the value is likely to be a GUID-style string of letters, numbers, and hyphens which changes each time you close and reopen a pbix file.

The SSMS object explorer shows the name of your database in the object hierarchy tree. Pressing the F8 key in SSMS makes the object explorer window appear.

The value used for DatabaseID needs to match precisely and potentially some square brackets removed from the start/end if dragging the value from the SSMS object explorer.

```
<ClearCache xmlns="http://schemas.microsoft.com/analysisservices/2003/engine">
    <Object>
        <DatabaseID>75b87f1a-e6fe-4d66-b268-a2d4fe9177fc</DatabaseID>
    </Object>
</ClearCache>
```

Figure 13-6. *XMLA query to clear cache for the data model*

Once the clear cache query runs and is successful, you should see a result similar to Figure 13-7. If the clear cache query fails, read the error message carefully, but it is likely to be caused by an incorrect DatabaseID.

```
<return xmlns="urn:schemas-microsoft-com:xml-analysis">
    <root xmlns="urn:schemas-microsoft-com:xml-analysis:empty" />
</return>
```

Figure 13-7. *The result of a successful XMLA query to clear cache*

```
<ClearCache xmlns="http://schemas.microsoft.com/analysisservices/2003/engine">
    <Object>
        <DatabaseID>7134f63d-ad2e-471f-84c1-f1cbb4aaef72</DatabaseID>
    </Object>
</ClearCache>

GO  ⬅━━━━━━

EVALUATE ROW("Basic Sum",SUM('Fact Sale'[Quantity]))

GO  ⬅━━━━━━
```

Figure 13-8. *Using the GO keyword to batch multiple statements in an SSMS DAX query window*

My next query

The next query expands on the previous example but substitutes the hardcoded constant with some DAX that uses the Storage Engine. The updated query is a simple SUM over the Quantity column and gets shown in Listing 13-4. This unfiltered code generates a single value of 8,950,628.

Listing 13-4. A Simple DAX query showing the SUM over the entire Quantity column

```
EVALUATE
    ROW (
            "Sum of Quantity",
            SUM('Fact Sale'[Quantity])
            )
```

Logical Plan

A four-line Logical Plan for the query in Listing 13-4 is shown in Figure 13-9.

```
AddColumns: RelLogOp DependOnCols()() 0-0 RequiredCols(0)(''[Sum of Quantity])
       Sum_Vertipaq: ScaLogOp DependOnCols()() Integer DominantValue=BLANK
            Scan_Vertipaq: RelLogOp DependOnCols()() 0-102 RequiredCols(12)('Fact Sale'[Quantity])
            'Fact Sale'[Quantity]: ScaLogOp DependOnCols(12)('Fact Sale'[Quantity]) Integer DominantValue=NONE
```

Figure 13-9. *The Logical Plan for the query in Listing 13-4*

The top line of the Logical Plan has not changed from the example in Listing 13-2. This top line relates to the final output and represents the ROW function used in the query.

The remaining lines in the plan relate to the SUM function used in the query. The second line has an operator name of Sum_VertiPaq and is a ScaLogOp type. Operator properties common to all Logical ScaLogOp types include the DependsOnCol, DataType, and DominantValue properties.

DependsOnCols

The DependsOnCols property shows two sets of parentheses and is the same for both ScaLogOp and RelLogOp. These can sometimes be empty, as they are in line 2 of the current Logical Plan. When these parentheses get populated with values, the first set shows a comma-separated list of numbers that relate to the internal column IDs within the data model. This number has no relationship to the Column ID in the model. This is simply a number that is unique in the current query and identifies an instance of a column. The second set of parentheses shows a fully qualified text version of the same columns.

The column numbers can be useful to help identify multiple references to the same column in a query. An example where this might happen is in a cumulative total calculation where an inner and outer loop keeps a different track of the same column.

The parentheses are not always populated and sometimes show as ()(), meaning there is no dependency on any column.

Column range (RelLogOp only)

For RelLogOp operations only, an additional range of numbers may appear following the DependsOnCol property. The numbers refer to a range of columns that determine the minimal set of columns required for the next part of the query. This range may not appear in the plan if the query does not involve a relation.

The format for the column range is *<start column id> - <end column id>*. In the example shown in Figure 13-9, the column range specified for the RegLogOp is a pretty wide range of 0–102. This column range is narrowed down to a specific column in the RequiredCols operator property. The RequiredCols operator can come from DependsOnCols or the column range.

RequiredCols (RegLogOp Only)

The RequiredCols operator property specifies the actual columns required to satisfy the operation. The pattern is similar to the DependsOnCols operator property in that there are two sets of parentheses. The first set contains a comma-separated list of numbers that indicate an internal column identifier, while the second set shows the fully qualified name of the column that includes the table name and column name.

DataType (ScaLogOp only)

Following the DependsOnCols property, one of the six datatypes is listed to show the datatype of the value returned. This value can also be blank. Both ScaLogOp operations in the current example use the Integer data type.

DominantValue (ScaLogOp only)

The final property of the scalar logical operator is the DominantValue property, which indicates the plan can identify or filter the column down to a specific value or a small number of values. This property can be set to NONE or otherwise might show a specific value. When the property has a specific value set, the DAX engine may pick a plan that can be significantly faster than a property set to NONE. This property value can be a useful indicator to look for when analyzing a logical plan.

Physical Plan

A five-line Physical Plan gets generated by the query in Listing 13-4 and shown in Figure 13-10.

```
AddColumns: IterPhyOp LogOp=AddColumns IterCols(0)(''[Sum of Quantity])
      SingletonTable: IterPhyOp LogOp=AddColumns
      SpoolLookup: LookupPhyOp LogOp=Sum_Vertipaq Double #Records=1 #KeyCols=120 #ValueCols=1 DominantValue=BLANK
            ProjectionSpool<ProjectFusion<Copy>>: SpoolPhyOp #Records=1
                  Cache: IterPhyOp #FieldCols=0 #ValueCols=1
```

Figure 13-10. *The Physical Plan for the query in Listing 13-4*

The top line is the outermost operation that uses the AddColumns operator to output a final result. The operation referenced by the second line uses the IterPhyOp type, even though this query only returns a single row. The AddColumns operator would not change if the query were designed to return more rows.

The third line represents a call to the Storage Engine and, for this plan, explains how the engine intends to satisfy the requirement to perform a sum. The LogOp property for this operator specifies a Sum_VertiPaq operator that gets used. The Sum_VertiPaq operator type is good news regarding performance and means the work to generate the value needed can be done while the Storage Engine is reading the data. No additional work is needed in the Formula Engine to produce the correct result.

The final two operators provide a little detail on the subnodes that relate to the SpoolLookup operator.

An important operator property in Physical Plans is the #Records operator property. This operator property provides insight into the number of rows in a spool that are processed by an operation. Sometimes this value may be much higher than you expect and may indicate an opportunity to reorganize filtering within the query to help reduce redundant and unnecessary workload.

VertiPaq operators

When using a DAX Query Plan to help improve performance, pivotal operators to look for in the Logical and Physical Plans are the VertiPaq operators.

VertiPaq operators work with the VertiPaq engine to retrieve data, and certain operations can take place *during* the process of retrieving data from the database that provide a significant overall boost to the performance of a query. VertiPaq operations can also run in parallel in a multithreaded environment which can lead to potential performance gains.

Any work that cannot be satisfied by a VertiPaq operator is passed to the single-threaded Formula Engine to resolve. The more data processing that can take place while data is being retrieved from the database, the better the outcome for performance.

Listing 13-5. A partial list of VertiPaq logical operators

```
Scan_Vertipaq
GroupBy_Vertipaq
Filter_Vertipaq
Sum_Vertipaq
Min_Vertipaq
Max_Vertipaq
Count_Vertipaq
DistinctCount_Vertipaq
Average_Vertipaq
Stdev.S_Vertipaq
Stdev.P_Vertipaq
Var.S_Vertipaq
Var.P_Vertipaq
```

The Query Plan in Figure 13-9 shows on the second line that the Sum_VertiPaq operator is used. This operator indicates that values in the [Quantity] column can be added together as the data is retrieved from the VertiPaq engine, meaning the number is available as soon as the read operation is complete. There is no requirement for the Formula Engine to perform additional processing to satisfy this query.

The Sum_VertiPaq operation property also appears in the Physical Plan shown in line 3 of Figure 13-10 in the key-value pair of LogOp = Sum_VertiPaq.

VertiPaq Query Events

Two Profiler events recommended to be added to a Profiler trace are the *VertiPaq SE Query Cache Match* and *VertiPaq SE Query End* events. These events display T-SQL-like pseudo-queries when reviewing and provide helpful feedback for debugging. These events also report a value for the duration of each Storage Engine query in the Profiler trace output. The duration value indicates how long each Storage Engine scan takes and is in milliseconds. The events specifically report any activity involved in querying the VertiPaq Storage Engine.

There are two Profiler events captured per Storage Engine query. The text for both is similar but has a slightly different *EventSubClass*.

Figure 13-11 shows the Trace Properties window that has both events added. The two events exist in the *Query Processing* section.

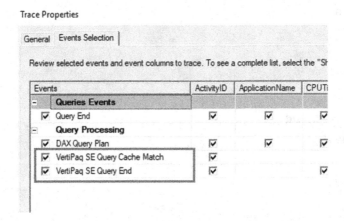

Figure 13-11. *Profiler events that track VertiPaq Storage Engine queries*

Consider the following DAX query in Listing 13-6 that builds a list of distinct dates using the 'Dimension Date' table and then uses the ADDCOLUMNS function to append a SUM value for each date.

Listing 13-6. A query to demonstrate DAX Profiler events

```
EVALUATE
    ADDCOLUMNS(
        VALUES('Dimension Date'[Date]) ,
        "Quantity Per Day",
        SUM('Fact Sale'[Quantity])
        )
```

EventClass	EventSubclass	Duration
DAX Query Plan	1 - DAX VertiPaq Logical Plan	
VertiPaq SE Query...	10 - Internal VertiPaq Scan	0
VertiPaq SE Query...	0 - VertiPaq Scan	0
VertiPaq SE Query...	10 - Internal VertiPaq Scan	0
VertiPaq SE Query...	0 - VertiPaq Scan	1
DAX Query Plan	2 - DAX VertiPaq Physical Plan	
Query End	3 - DAXQuery	14

Figure 13-12. *Profiler trace for the query in Listing 13-6*

The Profiler trace in Figure 13-12 shows seven lines in total. The top line has an EventClass of *DAX Query Plan* and relates to the Logical Plan. A simplified version of the Logical plan gets shown in Listing 13-7.

The Logical Plan suggests two VertiPaq scans take place during the query. These scans get referenced in lines 2 and 4 of the Logical Plan, and each line begins with the text *Scan_VertiPaq*. Two VertiPaq operations are required, first to generate a distinct list of date values from the 'Dimension Date' table and the second scan dealing with the Quantity column from the 'Fact Sale' table and belonging to the Sum_VertiPaq operation.

The operator name of the third line is *Sum_VertiPaq* which means the process of adding values together can take place while data gets retrieved from the underlying model.

Listing 13-7. Simplified Logical Plan for the query in Listing 13-6

```
AddColumns: RelLogOp ([Date], [Quantity Per Day])
    Scan_Vertipaq: RelLogOp ('Dimension Date'[Date])
    Sum_Vertipaq: ScaLogOp
        Scan_Vertipaq: RelLogOp ('Fact Sale'[Quantity])
        'Fact Sale'[Quantity]: ScaLogOp ([Quantity])
```

The Profiler Trace in Figure 13-11 shows four events in the *VertiPaq SE Query End* EventClass. These events represent two VertiPaq scans as a single scan will generate two events.

The sixth line of the Profiler Trace in Figure 13-11 shows a second *DAX Query Plan* event. The EventSubClass for this event is *2 – DAX VertiPaq Physical Plan* and represents the Physical Plan for the query. A simplified version of the Physical Plan is shown in Listing 13-8 that has some text omitted, but still preserves the intent.

Listing 13-8. Simplified Physical Plan for the query in Listing 13-6

```
AddColumns: IterPhyOp (0,1)([Date],[Quantity Per Day])
    Spool_Iterator: IterPhyOp LogOp=Scan_Vertipaq [Date]
        ProjectionSpool: SpoolPhyOp #Records=1461
            Cache: IterPhyOp
SpoolLookup: LookupPhyOp LogOp=Sum_Vertipaq
        ProjectionSpool: SpoolPhyOp #Records=1
            Cache: IterPhyOp
```

The outermost operation in the Physical Plan is the top row and relates to the ADDCOLUMNS function in the DAX query. This operation has a physical operator type of IterPhyOp that kicks off an iterator over the [Date] column. The #Records=1461 operator property specifies that there are 1,461 items to be iterated over and uses the logical operation of Scan_VertiPaq.

The fifth line represents a SpoolLookup operation. The physical operation type, in this case, is a LookupPhyOp which returns a single scalar value and uses the logical operation of Sum_VertiPaq. This line is the operation that takes care of adding all the values in the 'Fact Sale'[Quantity] column.

A rough translation of the Physical Plan is this: Perform a loop for every distinct date value found in the 'Dimension Date' table and then, for every date found, find a single value that is the sum total of the [Quantity] column in the 'Fact Sale' table. The order in which dates get processed is not essential. No single date relies on the output of another date. The task could be broken down into multiple parallel streams, and the result would still be the same.

The final line in the Profiler Trace is the *Query End* EventClass which provides a useful duration (in milliseconds) figure.

The results of the query in Listing 13-6 get shown in Figure 13-13. You may notice the value in the [Quantity Per Day] column does not change; rather, it reflects the overall total of the Quantity column in the 'Fact Sale' table over and over for each row. This omission is deliberate, and section "Context transition" later in this chapter shows how to identify and fix this using help from the Query Plans.

Dimension Date[Date]	[Quantity Per Day]
1/01/2013 12:00:00 AM	8950628
2/01/2013 12:00:00 AM	8950628
3/01/2013 12:00:00 AM	8950628
4/01/2013 12:00:00 AM	8950628
5/01/2013 12:00:00 AM	8950628
6/01/2013 12:00:00 AM	8950628
7/01/2013 12:00:00 AM	8950628
8/01/2013 12:00:00 AM	8950628

Figure 13-13. *Shows results from the query in Listing 13-6*

Plan optimization

As covered earlier in this chapter, when a query runs, several things happen including the generation of a Logical and Physical Plan to help determine the best approach to produce the required results.

The DAX Query Optimizer generates a Logical plan and tries to find what it thinks is the most efficient approach to complete the task. This optimization can mean queries written in multiple ways can end up with an identical plan and, therefore, identical performance characteristics.

Consider the following three *different* queries in Listing 13-9 that all produce the same result. All three DAX queries are preceded by a ClearCache block that gets abbreviated in this example for brevity.

Listing 13-9. Three different DAX queries with identical Logical and Physical plans

```
<ClearCache ... />
GO
EVALUATE
    ROW(
        "Basic SUM",
        SUM('Fact Sale'[Quantity])
        )
GO
<ClearCache .../>
GO
```

```
EVALUATE
    ROW(
        "Calculate SUM",
        CALCULATE(SUM('Fact Sale'[Quantity]))
        )
GO
<ClearCache .../>
GO
EVALUATE
    ROW(
        "SumX SUM",
        SUMX('Fact Sale','Fact Sale'[Quantity])
        )
```

The simplified Logical Plan, which is the same for all three queries, gets shown in Listing 13-10 and highlights a *Sum_VertiPaq* operator gets used for the query and results in optimal performance. The fact the Logical Plan is the same for all three queries also highlights that in simple cases, the SUMX function is no different from the SUM function or rather the SUM function is the same as the SUMX function.

The DAX Query Optimizer also has not used the Calculate logical operator for the middle query that uses the CALCULATE function. The Optimizer has determined that because no filters get supplied, a simpler and faster plan can get constructed. A different Logical and Physical plan gets generated if filter arguments get passed to the CALCULATE function.

Listing 13-10. Simplified logical plan for all three DAX queries in Listing 13-9

```
AddColumns: RelLogOp
    Sum_Vertipaq: ScaLogOp
        Scan_Vertipaq: RelLogOp 'Fact Sale'[Quantity]
        'Fact Sale'[Quantity]: ScaLogOp
```

The DAX Query Optimizer eliminates branches of logic from the Logical (and therefore Physical) Plan if the sections of the code are redundant. This optimization might be the case where only one output (THEN/ELSE) of an IF function is possible. The redundant operations get dropped from the plan.

The DAX Query Optimizer eliminates branches of logic from the Logical (and therefore Physical) Plan if it detects redundant code. This scenario may occur when only one output is possible from a conditional statement such as IF/ELSE. The redundant operations get dropped from the plan.

Flow control
Simple IF statement

The Query Optimizer creates a Logical Plan that only includes code paths deemed necessary to complete the query. If a DAX query includes an IF function with two possible outcomes, the Optimizer checks if both paths get used before finalizing the Logical Plan.

Consider the following query in Listing 13-11 that includes an IF function with unreachable code. The unreachable code is the ELSE section of the IF statement that includes a SUM function using the 'Fact Sale'[Profit] column.

Listing 13-11. An IF statement where the ELSE branch can never get reached

```
DEFINE
    VAR varTest = "Quantity"
EVALUATE
    ADDCOLUMNS(
        VALUES('Dimension Date'[Calendar Year]) ,
        "IF Test" ,
        IF (
            varTest = "Quantity" ,
            SUM('Fact Sale'[Quantity]) ,
            SUM('Fact Sale'[Profit]))    --<= redundant
        )
```

Listing 13-12. Simplified Logical Plan for the DAX query in Listing 13-11

```
varTest: Constant: ScaLogOp VarName=varTest
AddColumns: RelLogOp
    Scan_Vertipaq: RelLogOp [Calendar Year]
    Integer->Double: ScaLogOp DependOnCols [Calendar Year]
        Sum_Vertipaq: ScaLogOp MeasureRef=[Sum Quantity]
            Scan_Vertipaq: RelLogOp
            'Fact Sale'[Quantity]: ScaLogOp [Quantity]
```

The clear Logical plan in Listing 13-12 shows two VertiPaq scans. The first VertiPaq scan relates to the VALUES function and obtains the distinct list of values in the [Calendar Year] column. There are only four values returned.

Then for each of these values, a second VertiPaq scan using the *Sum_VertiPaq* operator queries the [Quantity] column. There is no reference in the Logical Plan to the 'Fact Sale'[Profit] column, even though the column gets directly referenced in the DAX query. The column is missing from the Logical Plan because the Query Optimizer has determined the ELSE path if the IF statement can never get reached.

If the DAX query in Listing 13-11 got modified so that the value assigned to the varTest variable was something other than 'Quantity', the DAX Query Optimizer would generate the same plan, only swapping the [Quantity] column with the [Profit] column.

An item of interest in the Logical Plan is the conversion from Integer to Double along the way. The conversion relates explicitly to the 'Fact Sale'[Quantity] column set to datatype Integer, and the Logical Plan determines an operation is required to convert this to decimal (Double) as part of the query.

Complex IF statement

If both code paths of an IF function are valid in a DAX query, the Optimizer includes additional operations to cover both paths in the Logical Plan. A slightly modified query designed to return a different measure when the year is equal to 2015 than for other years gets shown in Listing 13-13. The purpose of this query is to show how the Query Optimizer constructs a Logical Plan when both the THEN and ELSE paths of an IF function are valid.

Listing 13-13. Complex IF function with two valid code paths

```
DEFINE
    MEASURE 'Fact Sale'[Sum Quantity]
                = SUM('Fact Sale'[Quantity])
    MEASURE 'Fact Sale'[Sum Profit]
                = SUM('Fact Sale'[Profit])
EVALUATE
    ADDCOLUMNS(
        VALUES('Dimension Date'[Calendar Year]) ,
        "IF Test" ,
        IF (
            [Calendar Year] = 2015 ,
            [Sum Quantity]   ,      --<= Then
            [Sum Profit]     --<= Else
            )
        )
```

Listing 13-14. Simplified Logical Plan for the DAX query in Listing 13-13

```
AddColumns: RelLogOp
    Scan_Vertipaq: RelLogOp [Calendar Year])
    If: ScaLogOp 'Dimension Date'[Calendar Year]
        'Dimension Date'[Calendar Year] = 2015: ScaLogOp
        PredicateCheck: RelLogOp
            ScalarVarProxy: ScaLogOp
        PredicateCheck: RelLogOp
            ScalarVarProxy: ScaLogOp
        Sum_Vertipaq: ScaLogOp MeasureRef=[Sum Quantity]
            Scan_Vertipaq: RelLogOp
            'Fact Sale'[Quantity]: ScaLogOp
        Sum_Vertipaq: ScaLogOp MeasureRef=[Sum Profit]
            Scan_Vertipaq: RelLogOp
            'Fact Sale'[Profit]: ScaLogOp
```

The difference between the Logical Plan shown in Listing 13-14 and the Logical Plan shown in Listing 13-12 is the inclusion of an IF operator that includes multiple child nodes to cover the different code branches. The Logical Plan also indicates that three storage engine scans take place, with one scan focusing on retrieving a list of values in the [Calendar Year] column, while the other two use the *Sum_VertiPaq* operation type for speedy processing of the numbers depending on the conditional branch of the IF statement.

SWITCH statement

If the DAX query in Listing 13-13 was modified to use a SWITCH function in place of the IF function, the Query Optimizer would generate an identical plan.

However, something to be wary of when using the SWITCH function is that in some circumstances, all branches of a code path get evaluated even if it seems only one is possible. This unwanted situation may happen when a report uses a measure table to provide the end user with a dynamic experience that allows them to select a metric from a slicer. Consider the DAX calculated measure in Listing 13-15 that assigns to the *UserSelection* variable a selection made on a slicer over a column in a table called 'Measure Table.'

The expectation here is that when a user makes a single selection, the SWITCH statement only makes a call to a specific measure and only one of the three measures ever needs to be evaluated during a calculation.

Listing 13-15. Example of the SWITCH statement in a dynamic measure

```
SWITCH Measure =
VAR UserSelection = SELECTEDVALUE('Measure Table'[Column1])
RETURN
    SWITCH(
        TRUE() ,
        UserSelection = "Measure 1" , [Test Measure 1],
        UserSelection = "Measure 2" , [Test Measure 2],
        [Test Measure 3]
        )
```

CALCULATE statements

This next section focuses on interesting aspects of the query plan that relate to the usage of the CALCULATE function. The first section highlights how you might spot context transition within a query plan, while the second part provides an interesting example where you see significant performance gains by rewriting a slow-performing query to use the CALCULATE function.

Context transition

Consider the query in Listing 13-16 that returns a result with one row per [Calendar Year], along with an additional column with a value showing the SUM of the 'Fact Sale'[Quantity] column. In this example, the value in the [Quantity Per Year] column is the same value over and over and does not take into account the calendar year showing in the same row of the first column. The full result for the code at Listing 13-16 is shown in Figure 13-14.

Listing 13-16. Simple SUMX calculation without context transition

```
EVALUATE
    ADDCOLUMNS(
        VALUES('Dimension Date'[Calendar Year]) ,
        "Quantity Per year",
        SUMX('Fact Sale','Fact Sale'[Quantity])
        )
```

Dimension Date[Calendar Year]	[Quantity Per year]
2013	8950628
2014	8950628
2015	8950628
2016	8950628

Figure 13-14. *Shows resultset for the query in Listing 13-16*

Now consider a slightly modified version of the same query from Listing 13-16. The new query is in Listing 13-17, and the only difference is a CALCULATE function gets added around the SUMX function. No other modifications get made to the query, and no filter arguments get added to the CALCULATE function.

This modification enables context transition to take place and means row-level filters for the current row transition to the new context created by the CALCULATE statement.

Listing 13-17. Simple SUMX calculation with context transition taking place

```
EVALUATE
    ADDCOLUMNS(
            VALUES('Dimension Date'[Calendar Year]) ,
            "Quantity Per year" ,
            CALCULATE(
                    SUMX('Fact Sale','Fact Sale'[Quantity])
                    )
            )
```

Dimension Date[Calendar Year]	[Quantity Per year]
2013	2401657
2014	2567401
2015	2740266
2016	1241304

Figure 13-15. *Shows resultset for the query in Listing 13-17*

The simplified Logical Plan for the query in Listing 13-16 that does not use the CALCULATE function gets shown in Listing 13-18. For direct comparison, the simplified Logical Plan for the modified query in Listing 13-17 that uses the CALCULATE function gets shown in Listing 13-19. Figure 13-15 shows the full result for the code at Listing 13-17.

Listing 13-18. Simplified Logical Plan for the query in Listing 13-16 (no CALCULATE)

```
AddColumns: RelLogOp
    Scan_Vertipaq: RelLogOp
    Sum_Vertipaq: ScaLogOp
                  ↳ DependOnCols()()
        Scan_Vertipaq: RelLogOp
                      ↳ DependOnCols()()
                      ↳ 1-110
                      ↳ RequiredCols(13)('Fact Sale'[Quantity])
        'Fact Sale'[Quantity]: ScaLogOp
```

Listing 13-19. Simplified Logical Plan for the query in Listing 13-17 (with CALCULATE)

```
AddColumns: RelLogOp
    Scan_Vertipaq: RelLogOp
    Sum_Vertipaq: ScaLogOp
                    ↳ DependOnCols(0)([Calendar Year])
            Scan_Vertipaq: RelLogOp
                        ↳ DependOnCols(0)([Calendar Year])
                        ↳ 1-110
                        ↳ RequiredCols(0, 13)
                        ↳([Calendar Year], [Quantity])
            'Fact Sale'[Quantity]: ScaLogOp
```

For both simplified plans shown in Listings 13-16 (without CALCULATE) and Listing 13-17 (with CALCULATE), nonessential and identical operator properties get removed for brevity.

Both plans have five operations and therefore have five lines. The ↳ character gets used to highlight a break introduced to the line for readability.

The critical difference between the two plans is the operator property called DependsOnCols does not get populated with column values in the query that does not use the CALCULATE function.

However, in the version that uses CALCULATE, the DependsOnCols operator property provides a link between the two scans that take place. In both queries, two VertiPaq scans get executed; however, the Sum_VertiPaq operation takes into account the relevant [Calendar Year].

The Physical plans for both queries are very similar except for a line in the plan which relates to the Sum_VertiPaq logical operation that includes a LookupCols property that creates a link between the two scans.

Regarding performance, the two queries had similar performance characteristics over the 228,265-row 'Fact Sale' table but might be different with a much larger dataset.

Running total

Sticking with the CALCULATE function, another fun example to highlight using the Query Plan is a cumulative total calculation. Listing 13-20 shows two DAX queries written differently, but produce the same result.

Listing 13-20. Two DAX calculations that produce a running total over the [Quantity] column

```
<ClearCache .../>
GO
EVALUATE
    ADDCOLUMNS(
          VALUES('Dimension Date'[Date]),
          "Running total of Quantity" ,
          CALCULATE(
                SUMX(
                      'Fact Sale',
                      'Fact Sale'[Quantity]
                      ),
                'Dimension Date'[Date]<=
                      ↳ EARLIER('Dimension Date'[Date])
                )
          )
GO
<ClearCache .../>
GO
EVALUATE
    ADDCOLUMNS(
          VALUES('Dimension Date'[Date]),
          "Running total of Quantity" ,
          SUMX(
                FILTER(
                      'Fact Sale',
                      'Fact Sale'[Invoice Date Key]<=
                            ↳ EARLIER('Dimension Date'[Date])
                      ),
                'Fact Sale'[Quantity]
                )
          )
```

Both queries in the batch use a SUMX function in conjunction with the ADDCOLUMNS function to add a column showing a cumulative total for each day returned by the VALUES function.

Both queries return the same 1,461-row resultset. The same filter criteria also get used by both queries; however, the position and method of the filter make a big difference to the performance characteristics of each approach.

The first query in the example in Listing 13-20 uses a CALCULATE function to execute the core SUMX function and separates the filter logic to a filter parameter of the CALCULATE function, while the second query uses a FILTER function inside the SUMX expression.

Both queries perform two VertiPaq storage engine scans. The first VertiPaq scan relates to the VALUES function and determines the number of unique dates for the resultset, while the other VertiPaq scan uses the [Quantity] column.

The first query uses the CALCULATE function and reliably takes less than 1 second to complete. The second query which embeds filtering logic inside the SUMX expression takes anywhere between 20 and 30 seconds. This difference in performance is significant for what otherwise looks like two very similar queries.

The Query Plan for the two queries reveals clues as to the difference, and simplified versions of the two Logical Plans get shown in Listing 13-21 and Listing 13-22.

Listing 13-21. Simplified Logical Plan for the first query in Listing 13-20 (less than 1 second)

```
AddColumns: RelLogOp
    Scan_Vertipaq: RelLogOp
    Calculate: ScaLogOp
        Sum_Vertipaq: ScaLogOp
            Scan_Vertipaq: RelLogOp
            'Fact Sale'[Quantity]: ScaLogOp
        Filter: RelLogOp DependOnCols(0)
            Scan_Vertipaq: RelLogOp
            LessThanOrEqualTo: ScaLogOp
                'Dimension Date'[Date]: ScaLogOp
                'Dimension Date'[Date]: ScaLogOp
```

Listing 13-22. Simplified Logical Plan for second query in Listing 13-20 (20–30 seconds)

```
AddColumns: RelLogOp
    Scan_Vertipaq: RelLogOp
    SumX: ScaLogOp DependOnCols(0)
        Filter: RelLogOp DependOnCols(0)
            Scan_Vertipaq: RelLogOp
            LessThanOrEqualTo: ScaLogOp
                    'Fact Sale'[Invoice Date Key]: ScaLogOp
                    'Dimension Date'[Date]: ScaLogOp
        'Fact Sale'[Quantity]: ScaLogOp
```

Once again, noise and duplication have been removed to simplify the Logical Plans for readability. The big difference between the two plans is the first plan in Listing 13-21, which is the faster query, makes use of the *Sum_VertiPaq* operator, meaning the SUMX function can perform its work in the storage engine as the data gets read. The operation can potentially happen in parallel via multithreading. The use of Sum_VertiPaq here also explains why the overall query time is a little more than the time reported by the VertiPaq SE scan events.

The second Logical Plan in Listing 13-22 uses the SumX operator (at line 3) instead of the *Sum_VertiPaq* operator. The SumX operator does not run as part of the work performed by the Storage Engine and must wait until all the data is retrieved before it can proceed to fill the requirement using a single-threaded approach.

The Duration column in Figure 13-16 shows the duration profile for each query. Note that one of the VertiPaq SE queries in the second query is longer at 58 ms. This longer duration is because the scan injects an additional row identifier column needed by the SumX operation in the Formula Engine to help complete the work.

If you subtract the duration for the VertiPaq SE query events from the duration relating to the Query End event, you get an approximate figure that represents how much time was spent in the Formula Engine to satisfy the query.

Figure 13-16 shows the profiler trace details for the two queries in Listing 13-20. The top block of seven rows labelled 1 relates to the faster query that uses a CALCULATE function. The overall time taken for this query is shown in the Duration column of the Query End event and is 580 ms.

The bottom block of seven rows labelled 2 relates to the slower query that does not use a CALCULATE function. This query took 23 seconds to complete (23,251 ms) and did most of its work in the Formula Engine.

EventClass	EventSubclass	Duration
DAX Query Plan ①	1 - DAX VertiPaq Logical Plan	
VertiPaq SE Query End	10 - Internal VertiPaq Scan	1
VertiPaq SE Query End	0 - VertiPaq Scan	1
VertiPaq SE Query End	10 - Internal VertiPaq Scan	3
VertiPaq SE Query End	0 - VertiPaq Scan	3
DAX Query Plan	2 - DAX VertiPaq Physical Plan	
Query End	3 - DAXQuery	580
DAX Query Plan ②	1 - DAX VertiPaq Logical Plan	
VertiPaq SE Query End	10 - Internal VertiPaq Scan	58
VertiPaq SE Query End	0 - VertiPaq Scan	58
VertiPaq SE Query End	10 - Internal VertiPaq Scan	0
VertiPaq SE Query End	0 - VertiPaq Scan	0
DAX Query Plan	2 - DAX VertiPaq Physical Plan	
Query End	3 - DAXQuery	23251

Figure 13-16. *Shows Profiler trace for the query in Listing 13-20*

DAX Studio

The beautiful (and free) DAX Studio is an excellent companion when writing DAX queries, and among the many great features (as highlighted in Figure 13-17) included are the following:

(1) The ability to automate a ClearCache before each run

(2) Toggling on/off the Query Plan and Server timings including Scan and Cache timing

(3) The duration and CPU timings for any Storage Engine work (one line per scan)

(4) The totals for SE activity which can be used to derive Formula Engine times

(5) Tabs to show query results, Logical and Physical Plans, and the Timing details

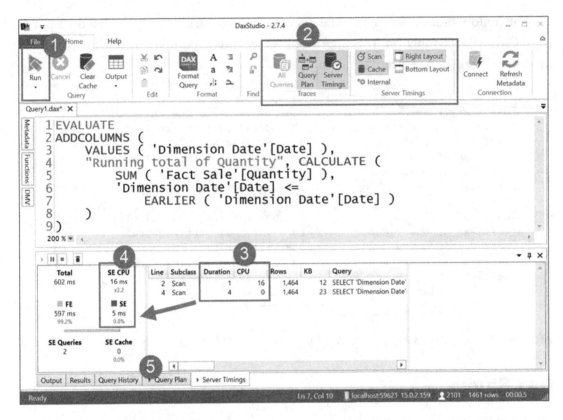

Figure 13-17. *DAX Studio showing the Server Timing information of a query*

The Query Plan tab shown at 5 in Figure 13-17 provides access to the Logical and Physical query plan when the Query Plan option gets enabled in the ribbon, at 2. This screen shows detail for the specific query executed from the query window. This detail includes extra rows at the top of the Physical Plan if the Clear Cache option gets enabled for the run.

Summary

This chapter has provided you with an introduction to the DAX Query Plan and how you might use it to help understand why specific queries run slowly.

The Query Plan combined with the Storage Engine profiler events can provide useful insights into the performance characteristics of various DAX queries.

When studying a plan, look for the existence of the various VertiPaq operators such as Sum_VertiPaq. The DAX engine runs at its fastest when running in pure VertiPaq mode.

When looking at the Physical Plan, look at the order and position of filters and nodes and don't be afraid to rewrite your query a different way so you can compare plans. The #Records operator property is also a very useful metric when optimizing queries for performance.

Always clear the cache when debugging for performance!

The only numbers provided for duration are for the DAX Query End event and the VertiPaq SE events.

For additional reading, I recommend an excellent whitepaper written by Alberto Ferrari in 2013 called "Understanding DAX Query Plans," along with anything written on the subject by Jeffrey Wang (the father of DAX) on his mdxdax.blogspot.com web site.

Scale Your Models

Introduction

There are two use cases to compose a DAX statement. The first one is to answer a specific question for a given dataset. The second one is to find an answer for a general question; this solution will allow us to answer similar questions much faster in the future.

For this reason, the latter solutions are also called patterns. At first glance, it seems foolish not to use an existing pattern or to develop a new one. But of course, sometimes it's not that easy because there is no time for the development of a pattern or just for doing a quick Internet search to find an already existing pattern and then just adapt this pattern to a given specific question. This seems odd, but developing a pattern is only possible if some if not all details of a system are known, especially how different situations will impact the intricate workings of all the components of the system. Developing these skills is costly (from a timing perspective). Admittedly, it is also a great feeling to find the answer to a question NOW; sometimes a solution has to be found now because the visual that is the most important visual used in a presentation this afternoon does not show the correct (or at least expected) values. This firefighting-like DAX development can be fun and rewarding in many ways, emotionally and intellectually. But as all real firefighters know, there is a risk. Fortunately, this risk in firefighting-like DAX development is only this: It's not possible to find a solution in a given time frame. As a firefighter enters unknown buildings, it's necessary to find some similar analogies. For this, let's consider the unknown building being the Power BI data model. A firefighter does not know how different rooms (the tables) are connected by doors (the relationships) and if the doors are one way or can be used in each direction (cross filter direction equals one to many or both).

This chapter is about how to become a better firefighter by providing techniques which can be used to scale a small data model to a large data model. The scaling of the data model will be done in DAX; this means preparing a scalable data model also trains the understanding of DAX in general and, of course, the interaction between DAX and the data model.

© Philip Seamark, Thomas Martens 2019
P. Seamark and T. Martens, *Pro DAX with Power BI*, https://doi.org/10.1007/978-1-4842-4897-3_14

Starting with a small data model allows to easily overlook the result and control the DAX statement with mental arithmetic. The scaling of the smaller dataset allows to check the fitness of the DAX statement. The fitness of the DAX statement compares to the mental and physical fitness of the firefighter. As mentioned throughout the book, the storage engine is the component of the VertiPaq engine that executes certain parts of the DAX statement very fast, whereas the formula engine can compute complex statements. The utilization of both components is essential for the success of the measure, particularly if the measure will be applied to a large data model.

Note   Scaling of a small data model does not mean to create a Big Data dataset that spans multiple petabytes. Scaling means to create a data model from tens or hundreds of rows to thousands or millions.

Biased data

Nowadays, it's not good to use a biased dataset, at least not if the dataset originates from a thoughtless experiment. To better understand biased data, imagine a survey is conducted to predict if a new product that will be launched in the near future will be successful. If this product is a smartphone with some extra features for the older people, you have to make sure that the survey not only considers kids from the high school in your neighborhood, as these kids will rate the features as they are not members of the target group.

Nevertheless, it's also not a good idea to create a dataset that originates from an algorithm that is entirely randomized. This is because data is "biased" due to the following reasons:

- Some products are favored by customers, and for this reason, they appear more often in a sales dataset.

- Some products cannot be sold to some markets due to some legal constraints; for this, these products will not appear in the dataset if geospatial-related analysis is performed upon the dataset.

- Some products will not appear across the complete time frame as these products have been launched at a later point in time or have been vanished as their product life came to an end.

There are a lot of reasons why the dataset that is used to develop a measure will be biased. The main reason is simply this: A biased dataset created by an algorithm is more similar to a real-life dataset than a completely randomized dataset.

Dimension and fact tables

Dimension and fact tables are the core of the data model; if these tables do not contain the relevant data, no DAX statement will be able to calculate just the most straightforward measure. The task to design a scalable data model that contains sample data is a challenging task. For this reason, it's necessary to keep in mind that the model has to be simple, simple enough to do mental arithmetic, but complex enough to validate the result of a measure under different circumstances, meaning using the measure in various visuals where implicit filters will impact the number of rows that are aggregated. Assuming that a measure calculates the total quantity at any given point in time without the consideration of any other dimension (besides the calendar dimension), it's necessary to test the measure for returning the same value if a column from a different table is used to filter the fact table, for example, a column from the customer table.

A great reference about different types of dimension and fact tables is the book *Star Schema: The Complete Reference* by Christopher Adamson. This book can be used to create various data models that represent various business processes.

The focus of this chapter is not to create a complete data model for a specific business process. Here, it's only about the creation of some tables that can be used to test a measure for correctness and if the measure will perform on a larger dataset.

The sample data model is created by a configuration table that, of course, is defined by using a DAX statement. Listing 14-1 shows the DAX statement that defines this configuration table.

Listing 14-1. The configuration table

```
Sample Data Configuration =
UNION(
    ROW("Parameter Name" , "No Of Products to use", "Parameter Value" , 1000)
    , ROW("Parameter Name" , "No Of Events", "Parameter Value" , 10)
```

```
    , ROW("Parameter Name" , "No of Products per events (max)" , "Paramter
    Value" , 5)
    , ROW("Parameter Name" , "No of days ahead" , "Paramter Value" , 10)
)
```

Different parameters are stored in a table that can be enhanced when needed.

The dimension table

Neglecting any business-related content, a dimension table can be reduced to the following:

- A single row uniquely identifies a business object, for example, a product.

- Attributes are used to describe the business object, for example, the product color.

Note   Throughout this chapter, the Power BI file "CH14 – the sample model.pbix" is used. This pbix file is using the Excel file "Dimension Template.xlsx" as a data source. Be aware that the data from this file even contained in the data model is not used in the final data model, meaning there is no relationship to this model.

As this pbix file is using an Excel file as a data source, the path to this Excel file is configured as a parameter that can be changed very quickly to adopt the file to your file location. Query parameters are an old feature (if this could ever be possible for a product that is as young as Power BI is). Here is a link to the announcement of query parameters: https://powerbi.microsoft.com/en-us/blog/deep-dive-into-query-parameters-and-power-bi-templates/.

The Excel source file "Dimension Template.xlsx"

If you are wondering why an Excel source file is used just to store some information and why not only use DAX even to store something simple as the product color, the answer to this question is straightforward. It's much more comfortable to adopt the Excel file to some of your specific needs than adjusting DAX code to store myriads of liters like the names of colors.

The dimension base table

When creating a sample data model, it's necessary to make sure that a dimension table does not contain any duplicate identifier. Listing 14-2 shows the DAX statement that creates and populates the table Dimension Product - base table; the listing just creates a base table containing an ID column and a name column. It's necessary to notice that by no means the dimension created has to be a product dimension. Similar DAX code can be used to create a customer dimension.

Listing 14-2. Product dimension – base table

```
Dimension Product - base table =
var NoOfProductsToUse = [_No of Products To Use] --Maybe you will not
receive the exact number as duplicate rows will be eliminated because of
the usage of DISTINCT
var NoOfProductsAvailable = COUNTROWS('Dimension Template')
var ZeroPadding = CEILING(LOG10(NoOfProductsAvailable),1)
var p0_productbasetable = DISTINCT(
        SELECTCOLUMNS(
            ADDCOLUMNS(
                GENERATESERIES(1 , NoOfProductsToUse , 1)
                , "Product ID" , RANDBETWEEN(1 , NoOfProductsAvailable)
            )
            , "Product ID" , [Product ID]
            , "Product Name" , "P " & RIGHT(REPT("0" , ZeroPadding)
            & [Product ID], LEN(NoOfProductsAvailable))
        )
    )
return
p0_productbasetable
```

The DAX statement creates a table where the number of rows created depends on the measure NoOfProductsToUse. This measure controls how many rows, meaning products, will be created by using the DAX function GENERATESERIES.

At this stage, the dimension table just contains two columns:

- Product ID

- Product Name

The Product ID is an integer value. The Product ID is created by using the DAX function RANDBETWEEN. RANDBETWEEN performs a random draw from the given range of integers with replacement. Each value of the range defined by RANDBETWEEN(1 , NoOfProductsAvailable) has the same chance to be drawn. Unfortunately, this also means that the second draw can once again choose the corresponding integer value. As it is necessary to create a list of unique integer values that will represent the Product ID, it's also required to use DISTINCT as the outermost function creating the base table. The DAX function SELECTCOLUMNS is used to select the column Product ID (a random integer value) and also creates Product Name (a simple text concatenation, mainly to create a string value). Another use of SELECTCOLUMNS is to omit the column value that is automatically created by GENERATESERIES. As you start using and changing the parameters available, you will encounter situations where the parameter No Of Products to use and the number of product rows created will not match, as it is shown in Figure 14-1.

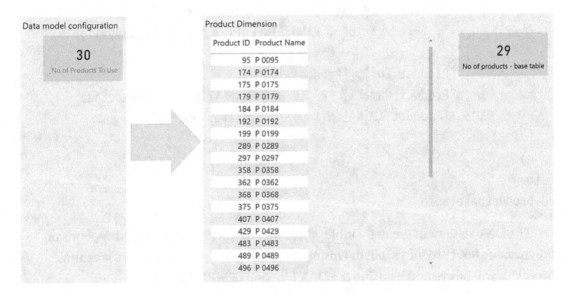

Figure 14-1. *Something unexpected*

Caution   Throughout this chapter, IDs will be used, as numeric data types will be aggregated automatically using the SUM aggregation if used on visuals. For this reason, make sure that the aggregation is adjusted to "Don't summarize."

The expectation – the number of products used in the table `Dimension Product - base table` will equal 30, by using the measure from Listing 14-3.

Listing 14-3. No of products – base table

```
No of products - base table =
COUNTROWS('Dimension Product - base table')
```

Sometimes the measure will not return what has been configured.

It's necessary to always keep in mind that the products that will be used have been chosen randomly. For this reason, as already mentioned, it can happen that a `Product ID` will be "drawn" more than once. Using the function `DISTINCT` will correct this "error." And for this reason, it can be possible that both measures will show different results.

Binning and groupings to create hierarchies

Next to creating sample data, it's necessary to create hierarchies. This is simply due to the fact that hierarchies group the unique values of a dimension into chunks. Products will be grouped into Product Subcategories, and Product Subcategories will be grouped into Product Categories.

Creating hierarchies can be a tedious piece of work. To omit this tedious and error-prone work, an algorithm is used to create a hierarchy. The grouping of IDs by using algorithms can be decomposed into two components:

- Determine the number of groups (bins).

- Assign an ID to a group.

As business objects are very often grouped into more than one hierarchy, two different algorithms are used. More information about the used algorithms can be found here: `https://en.wikipedia.org/wiki/Histogram`.

As it's not that much important what algorithms are used, the more simple ones are used:

- Sturges' formula

- Rice rule

Listing 14-4 shows the complete DAX statement to create the table Product Dimension – hierarchy. This table contains two hierarchies:

- Product ID ➤ SG1 ... (Subgroup 1) ➤ G1 ... (Group 1)

- Product ID ➤ SG2 ... (Subgroup 2) ➤ G2 ... (Group 2)

The essential parts are highlighted and explained in more detail after the listing.

Listing 14-4. Product dimension – hierarchy

```
Dimension Product - hierarchy =
var NoOfProductsToUse = [_No of Products To Use] --Maybe you will not
receive the exact number as duplicate rows will be eliminated because of
the usage of DISTINCT
var NoOfProductsAvailable = COUNTROWS('Dimension Template')
var ZeroPadding = CEILING(LOG10(NoOfProductsAvailable),1)
var p0_productbasetable = DISTINCT(
        SELECTCOLUMNS(
            ADDCOLUMNS(
                GENERATESERIES(1 , NoOfProductsToUse , 1)
                , "Product ID" , RANDBETWEEN(1 , NoOfProductsAvailable)
            )
            , "Product ID" , [Product ID]
            , "Product Name" , "P " & RIGHT(REPT("0" , ZeroPadding)
            & [Product ID], LEN(NoOfProductsAvailable))
        )
    )

/* ************************************************************** */
/* general information used for the binning to create hierarchies */
var NoOfRowsBaseTable = COUNTROWS(p0_productbasetable)
```

```
var MaxMinRange = MAXX(p0_productbasetable,[Product ID]) - MINX(p0_
productbasetable,[Product ID])

/* ************************************************************* */
/* using Sturges Formula to determine the bin width */
var noOfGroupsSturges = ROUND(LOG(NoOfRowsBaseTable,2),0)+1
var subgroupWidthSturges = CEILING((MaxMinRange/noOfGroupsSturges),1)
var groupWidthSturges = subgroupWidthSturges * 2

/* ************************************************************* */
/* using Rice Rule to determine the bin width */
var noOfGroupsRice = ROUND(2 * POWER(COUNTROWS(p0_productbasetable),1/3),0)
var subgroupWidthRice = CEILING((MaxMinRange/noOfGroupsRice),1)
var groupWidthRice = subgroupWidthRice * 2

var p1_productsWithHiearchy =
    ADDCOLUMNS(
        p0_productbasetable

        ,
        /* the formula to create the number of groups has been taken from:
        https://en.wikipedia.org/wiki/Histogram */
        /* ************************************************************* */
        /* using Sturges Formula to determine the bin width */
        "SG1 range"

            ,
            var groupNumber = CEILING([Product ID] / subgroupWidthSturges,1)
            return
            "SG1 range " & 1+(groupNumber-1) * subgroupWidthSturges & " to
            "  & FLOOR(groupNumber * subgroupWidthSturges,1)
        , "SG1 number"

            ,
            "SG1 number" & RIGHT(REPT("0", 3) & CEILING([Product ID] /
            subgroupWidthSturges,1),3)
        ,"G1 range"

            ,
            var groupNumber = CEILING([Product ID] / groupWidthSturges,1)
```

```
        return
        "G1 range " & 1+(groupNumber-1) * groupWidthSturges & " to "  &
        FLOOR(groupNumber * groupWidthSturges,1)
    , "G1 number"

        ,
        "G1 number " & RIGHT(REPT("0", 3) & CEILING([Product ID] /
        groupWidthSturges,1),3)

        /* ****************************************************** */
        /* using Rice Formula to determine the bin width */
    ,"SG2 range"

        ,
        var groupNumber = CEILING([Product ID] / subgroupWidthRice,1)
        return
        "SG2 range " & 1+(groupNumber-1) * subgroupWidthRice & " to
        "  & FLOOR(groupNumber * subgroupWidthRice,1)
    , "SG2 number"

        ,
        "SG2 number" & RIGHT(REPT("0", 3) & CEILING([Product ID] /
        subgroupWidthRice,1),3)
    ,"G2 range"

        ,
        var groupNumber = CEILING([Product ID] / groupWidthRice,1)
        return
        "G2 range " & 1+(groupNumber-1) * groupWidthRice & " to "  &
        FLOOR(groupNumber * groupWidthRice,1)
    , "G2 number"

        ,
        "G2 number " & RIGHT(REPT("0", 3) & CEILING([Product ID] /
        groupWidthRice,1),3)
    )
return
p1_productsWithHiearchy
```

The lines from Listing 14-4 boil down into three sections. These sections are used to store values that are used within the table iterator function ADDCOLUMNS that adds the columns to the Product Dimension – base table. The first section stores general information about the base table. Both algorithms use the same information.

```
************************************************************* */
/* general information used for the binning to create hierarchies */
var NoOfRowsBaseTable = COUNTROWS(pO_productbasetable)
var MaxMinRange = MAXX(pO_productbasetable,[Product ID]) - MINX(pO_
productbasetable,[Product ID])
```

The values that are stored are

- The number of rows in the base table.

- The range of product IDs. This range can be calculated by using the formula MAX(list) – MIN(list).

- As the list, namely, the Product ID column, is contained in a "virtual" table, the table iterator functions MAXX and MINX have to be used.

Both algorithms are using both the abovementioned values to determine the number of bins and the width of each bin. The next section just describes the calculation for Sturges' formula:

```
/* using Sturges Formula to determine the bin width */
var noOfGroupsSturges = ROUND(LOG(NoOfRowsBaseTable,2),0)+1
var subgroupWidthSturges = CEILING((MaxMinRange/noOfGroupsSturges),1)
var groupWidthSturges = subgroupWidthSturges * 2
```

The preceding lines are using Sturges' formula to calculate the number of bins. This value is stored to the variable noOfGroupsSturges. The width of each group is calculated and stored to the variable subgroupWidthSturges. Even if the same width is used for each group, this does not mean that calculation of the width does not consider the distribution of the IDs.

After the sections mentioned, there is a small thing left; the defined variables have to be used to create the additional columns that will represent the hierarchies. The next lines show how these variables are used for only two of the eight columns that will be created, as these lines will be repeated for all the other columns with just minor changes by using different variables:

```
"SG1 range"
        ,
        var groupNumber = CEILING([Product ID] / subgroupWidthSturges,1)
        return
        "SG1 range " & 1+(groupNumber-1) * subgroupWidthSturges & " to
        "  & FLOOR(groupNumber * subgroupWidthSturges,1)
, "SG1 number"
        ,
        "SG1 number" & RIGHT(REPT("0", 3) & CEILING([Product ID] /
        subgroupWidthSturges,1),3)
```

The column SG1 range contains the hierarchy level for the subgroup that is based on Sturges' formula; text concatenation is used to provide additional information about product IDs by adding the lower and upper bound of the IDs in that specific range. What happens is this:

- Store the group number for the Product ID of the current row from the base table to the variable groupNumber using CEILING.

- Concatenate the lower bound with the upper bound with a prefix and the word "to" for better readability.

Figure 14-2 shows some of the columns of the table Product Dimension – hierarchy, created with the parameter No Of Products to use = 30.

Product Dimension

Product Name	SG1 range	G1 range	SG2 range	G2 range
P 0016	SG1 range 1 to 163	G1 range 1 to 326	SG2 range 1 to 163	G2 range 1 to 326
P 0025	SG1 range 1 to 163	G1 range 1 to 326	SG2 range 1 to 163	G2 range 1 to 326
P 0027	SG1 range 1 to 163	G1 range 1 to 326	SG2 range 1 to 163	G2 range 1 to 326
P 0038	SG1 range 1 to 163	G1 range 1 to 326	SG2 range 1 to 163	G2 range 1 to 326
P 0167	SG1 range 164 to 326	G1 range 1 to 326	SG2 range 164 to 326	G2 range 1 to 326
P 0195	SG1 range 164 to 326	G1 range 1 to 326	SG2 range 164 to 326	G2 range 1 to 326
P 0214	SG1 range 164 to 326	G1 range 1 to 326	SG2 range 164 to 326	G2 range 1 to 326
P 0289	SG1 range 164 to 326	G1 range 1 to 326	SG2 range 164 to 326	G2 range 1 to 326
P 0380	SG1 range 327 to 489	G1 range 327 to 652	SG2 range 327 to 489	G2 range 327 to 652
P 0409	SG1 range 327 to 489	G1 range 327 to 652	SG2 range 327 to 489	G2 range 327 to 652
P 0440	SG1 range 327 to 489	G1 range 327 to 652	SG2 range 327 to 489	G2 range 327 to 652
P 0570	SG1 range 490 to 652	G1 range 327 to 652	SG2 range 490 to 652	G2 range 327 to 652
P 0588	SG1 range 490 to 652	G1 range 327 to 652	SG2 range 490 to 652	G2 range 327 to 652
P 0592	SG1 range 490 to 652	G1 range 327 to 652	SG2 range 490 to 652	G2 range 327 to 652
P 0631	SG1 range 490 to 652	G1 range 327 to 652	SG2 range 490 to 652	G2 range 327 to 652
P 0647	SG1 range 490 to 652	G1 range 327 to 652	SG2 range 490 to 652	G2 range 327 to 652
P 0709	SG1 range 653 to 815	G1 range 653 to 978	SG2 range 653 to 815	G2 range 653 to 978
P 0723	SG1 range 653 to 815	G1 range 653 to 978	SG2 range 653 to 815	G2 range 653 to 978

Figure 14-2. Small product dimension – hierarchies

Note   The figure may not match the result you are seeing, because of the random values that are used to create the content of this table.

If you watch closely, you will realize that both hierarchies SG1/G1 and SG2/G2 are identical. This is because of the similarities of both algorithms if a small number is used, as this is the case in the preceding example.

Figure 14-3 shows the same table but instead created with the parameter No Of Products to use = 100.

Product Dimension

Product Name	SG1 range	G1 range	SG2 range	G2 range
P 0007	SG1 range 1 to 120	G1 range 1 to 240	SG2 range 1 to 107	G2 range 1 to 214
P 0008	SG1 range 1 to 120	G1 range 1 to 240	SG2 range 1 to 107	G2 range 1 to 214
P 0034	SG1 range 1 to 120	G1 range 1 to 240	SG2 range 1 to 107	G2 range 1 to 214
P 0050	SG1 range 1 to 120	G1 range 1 to 240	SG2 range 1 to 107	G2 range 1 to 214
P 0055	SG1 range 1 to 120	G1 range 1 to 240	SG2 range 1 to 107	G2 range 1 to 214
P 0068	SG1 range 1 to 120	G1 range 1 to 240	SG2 range 1 to 107	G2 range 1 to 214
P 0069	SG1 range 1 to 120	G1 range 1 to 240	SG2 range 1 to 107	G2 range 1 to 214
P 0098	SG1 range 1 to 120	G1 range 1 to 240	SG2 range 1 to 107	G2 range 1 to 214
P 0100	SG1 range 1 to 120	G1 range 1 to 240	SG2 range 1 to 107	G2 range 1 to 214
P 0113	SG1 range 1 to 120	G1 range 1 to 240	SG2 range 108 to 214	G2 range 1 to 214
P 0121	SG1 range 121 to 240	G1 range 1 to 240	SG2 range 108 to 214	G2 range 1 to 214
P 0125	SG1 range 121 to 240	G1 range 1 to 240	SG2 range 108 to 214	G2 range 1 to 214
P 0132	SG1 range 121 to 240	G1 range 1 to 240	SG2 range 108 to 214	G2 range 1 to 214
P 0144	SG1 range 121 to 240	G1 range 1 to 240	SG2 range 108 to 214	G2 range 1 to 214
P 0148	SG1 range 121 to 240	G1 range 1 to 240	SG2 range 108 to 214	G2 range 1 to 214
P 0166	SG1 range 121 to 240	G1 range 1 to 240	SG2 range 108 to 214	G2 range 1 to 214
P 0167	SG1 range 121 to 240	G1 range 1 to 240	SG2 range 108 to 214	G2 range 1 to 214
P 0178	SG1 range 121 to 240	G1 range 1 to 240	SG2 range 108 to 214	G2 range 1 to 214

Figure 14-3. Larger product dimension – hierarchies

The difference of both algorithms will become much more evident if even more products are used to create a large dimension table (please be aware that large does not mean millions of rows in this example). Figure 14-4 shows the assignment of Product IDs with parameter No Of Products to use = 1000.

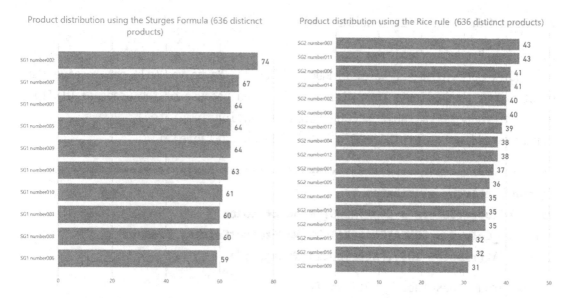

Figure 14-4. *Large product dimension – hierarchies distribution*

Other dimension attributes

Creating hierarchies is vital to be able to test measures thoroughly. But the attributes that are forming hierarchies are not the only attributes that will be used. For example, something simple as a product color is also a very important attribute. Of course, it's also possible to use any other attribute as an additional column in the product table. Attributes (columns) that are not used to form a hierarchy are a great helper to understand if measures are returning the correct value because a selected value from this kind of attributes is filtering out the number of products. This can have a great impact on any calculations that are trying to extract insights where the hierarchy plays a major part. As simple as this may sound, sometimes these simple things are not considered for testing a measure.

For this reason, this chapter is about adding a simple attribute like color to the table Product Dimension – color. This table is based on the table Product Dimension – base table. This is only for simplicity. At the end of this chapter, there is a reference to a final listing that combines all the different aspects into one DAX statement to create the final dimension table.

Listing 14-5 shows how the color attribute will be added to the dimension table. Once again, the important parts are highlighted and explained after the listing in more detail.

Listing 14-5. Product dimension – color

```
Dimension Product - color =
var NoOfProductsToUse = [_No of Products To Use] --Maybe you will not receive the
exact number as duplicate rows will be eliminated because of the usage of DISTINCT
var NoOfProductsAvailable = COUNTROWS('Dimension Template')
var ZeroPadding = CEILING(LOG10(NoOfProductsAvailable),1)

/* ************************************************************** */
/* non hierarchy attributes */
var NoOfroductColors = COUNTROWS('Product_Color')
var path_Colors = CONCATENATEX(TOPN(NoOfroductColors,'Product_
Color',[Attribute1_ID]),[Attribute1_Name],"|")

var p0_productbasetable = DISTINCT(
        SELECTCOLUMNS(
            ADDCOLUMNS(
                GENERATESERIES(1 , NoOfProductsToUse , 1)
                , "Product ID" , RANDBETWEEN(1 , NoOfProductsAvailable)
            )
            , "Product ID" , [Product ID]
            , "Product Name" , "P " & RIGHT(REPT("0" , ZeroPadding)
            & [Product ID], LEN(NoOfProductsAvailable))
        )
    )
var p2_productcolor =
ADDCOLUMNS(
    p0_productbasetable
    ,/* this is the product color */
    "Color"
        ,
        var randomvalue = RANDBETWEEN(1,NoOfroductColors)
        return
        PATHITEM(path_Colors ,randomvalue, TEXT)
)

return
p2_productcolor
```

The lines

```
/* non hierarchy attributes */
var NoOfproductColors = COUNTROWS('Product_Color')
var path_Colors = CONCATENATEX(TOPN(NoOfproductColors,'Product_
Color',[Attribute1_ID]),[Attribute1_Name],"|")
```

work like this:

- The number of colors is stored to the variable NoOfProductColors.

- TOPN returns an ordered table, ordered by the column Attribute1_
  ID. This table then is used to create a string of all color names using
  the column Attribute1_Name. Using the separator "|" allows using all
  the PATH functions like PATHITEM.

The lines

```
var randomvalue = RANDBETWEEN(1,NoOfproductColors)
      return
      PATHITEM(path_Colors ,randomvalue, TEXT)
```

create a

- Random integer using RANDBETWEEN. This integer is stored to the
  variable randomvalue.

The function PATHITEM allows retrieving an item (the color) from a string where
items are separated by "|."

As promised, Listing 14-6 will create the complete dimension table Product
Dimension - final.

Listing 14-6. product dimension - final

```
Dimension Product - final =
var NoOfProductsToUse = [_No of Products To Use] --Maybe you will not
receive the exact number as duplicate rows will be eliminated because of
the usage of DISTINCT
var NoOfProductsAvailable = COUNTROWS('Dimension Template')
var ZeroPadding = CEILING(LOG10(NoOfProductsAvailable),1)
```

```
var p0_productbasetable = DISTINCT(
        SELECTCOLUMNS(
            ADDCOLUMNS(
                GENERATESERIES(1 , NoOfProductsToUse , 1)
                , "Product ID" , RANDBETWEEN(1 , NoOfProductsAvailable)
            )
            , "Product ID" , [Product ID]
            , "Product Name" , "P " & RIGHT(REPT("0" , ZeroPadding)
            & [Product ID], LEN(NoOfProductsAvailable))
        )
    )

/* ************************************************************** */
/* general information used for the binning to create hierarchies */
var NoOfRowsBaseTable = COUNTROWS(p0_productbasetable)
var MaxMinRange = MAXX(p0_productbasetable,[Product ID]) - MINX(p0_
productbasetable,[Product ID])

/* ************************************************************** */
/* using Sturges Formula to determine the bin width */
var noOfGroupsSturges = ROUND(LOG(NoOfRowsBaseTable,2),0)+1
var subgroupWidthSturges = CEILING((MaxMinRange/noOfGroupsSturges),1)
var groupWidthSturges = subgroupWidthSturges * 2

/* ************************************************************** */
/* using Sturges Formula to determine the bin width */
var noOfGroupsRice = ROUND(2 * POWER(COUNTROWS(p0_productbasetable),1/3),0)
var subgroupWidthRice = CEILING((MaxMinRange/noOfGroupsRice),1)
var groupWidthRice = subgroupWidthRice * 2

/* ************************************************************** */
/* non hierarchy attributes */
var NoOfproductColors = COUNTROWS('Product_Color')
var path_Colors = CONCATENATEX(TOPN(NoOfproductColors,'Product_
Color',[Attribute1_ID]),[Attribute1_Name],"|")
```

```
var p1_productsWithHiearchy =
    ADDCOLUMNS(
        p0_productbasetable

        ,
        /* the formula to create the number of groups has been taken from:
        https://en.wikipedia.org/wiki/Histogram */
        /* ********************************************************* */
        /* using Sturges Formula to determine the bin width */
        "SG1 range"

            ,
            var groupNumber = CEILING([Product ID] / subgroupWidthSturges,1)
            return
            "SG1 range " & 1+(groupNumber-1) * subgroupWidthSturges & " to
            " & FLOOR(groupNumber * subgroupWidthSturges,1)
        , "SG1 number"

            ,
            "SG1 number" & RIGHT(REPT("0", 3) & CEILING([Product ID] /
            subgroupWidthSturges,1),3)
        ,"G1 range"

            ,
            var groupNumber = CEILING([Product ID] / groupWidthSturges,1)
            return
            "G1 range " & 1+(groupNumber-1) * groupWidthSturges & " to " &
            FLOOR(groupNumber * groupWidthSturges,1)
        , "G1 number"

            ,
            "G1 number " & RIGHT(REPT("0", 3) & CEILING([Product ID] /
            groupWidthSturges,1),3)

            /* ********************************************************* */
            /* using Rice Formula to determine the bin width */
        ,"SG2 range"

            ,
            var groupNumber = CEILING([Product ID] / subgroupWidthRice,1)
            return
```

```
            "SG2 range " & 1+(groupNumber-1) * subgroupWidthRice & " to
            " & FLOOR(groupNumber * subgroupWidthRice,1)
        , "SG2 number"

            ,
            "SG2 number" & RIGHT(REPT("0", 3) & CEILING([Product ID] /
            subgroupWidthRice,1),3)
        ,"G2 range"

            ,
            var groupNumber = CEILING([Product ID] / groupWidthRice,1)
            return
            "G2 range " & 1+(groupNumber-1) * groupWidthRice & " to "  &
            FLOOR(groupNumber * groupWidthRice,1)
        , "G2 number"

            ,
            "G2 number " & RIGHT(REPT("0", 3) & CEILING([Product ID] /
            groupWidthRice,1),3)
    )

var p2_final =
    ADDCOLUMNS(
        p1_productsWithHiearchy
        ,/* this is the product color */
        "Color"

        ,
        var randomvalue = RANDBETWEEN(1,NoOfproductColors)
        return
        PATHITEM(path_Colors ,randomvalue, TEXT)
    )
return
p2_final
```

The fact table

At its core, a fact table stores the measurement(s) of an event. Creating a scalable data model, this simple fact has its difficulties. For this reason, two typical fact tables will be created in this chapter.

Order and order line

Let's assume an event is a sale, meaning someone (a customer) is buying something (a product). This simple assumption comes with a lot of questions that we have to answer. Some of these questions are as follows:

- Can a product be sold more than once during an event?

  - The simple answer is yes. The measurement logs the product and the quantity, but there are situations where an event logs a product more than once, for example, buying products on the supermarket. Depending on the retailer, all products are scanned separately, but it also happens that a cashier scans a product just once and enters the quantity manually if more than one product is bought. Of course, as things happen, suddenly the same product surfaces underneath some other products and will be scanned once again.

We have to decide what's possible. As we draw randomly from a list of available products, we decide that a single event logs a product just once, but of course, it's possible that can be bought more than once.

Listing 14-7 shows a very simple fact that adheres to the preceding definition.

Listing 14-7. Fact table – simple

```
Fact - simple =
var ProductsPath = CONCATENATEX(TOPN([_No of Products per transaction
(max)], 'Dimension Product - hierarchy','Dimension Product -
hierarchy'[Product ID],ASC),[Product ID],"|")
return
ADDCOLUMNS(
    GENERATE(
        SELECTCOLUMNS(
                GENERATESERIES(1 , [_No of Events] , 1 )
            , "Order ID" , [Value]
        )
        ,var noofproductspertransaction = RANDBETWEEN(1 , [_No of Products
        per transaction (max)])
        return
```

```
    DISTINCT(
        SELECTCOLUMNS(
            ADDCOLUMNS(
                GENERATESERIES(1 ,  noofproductspertransaction)
                , "Product ID" , PATHITEM(ProductsPath , RANDBETWEEN(1
                , PATHLENGTH(ProductsPath)) , INTEGER)
            )
            , "Orderline ID" , [Value]
            , "Product ID" , [Product ID]
        )
    )
)
,"Quantity" , RANDBETWEEN(1 , 10)
)
```

This is what happens in Listing 14-7:

- An ordered list of products is stored to the variable ProducsPath. Once again this list is created by using the function CONCATENATEX with the special separator "|."

- The function GENERATE combines two nested tables created by GENERATESERIES, whereas the outer GENERATESERIES represents the event (the order) and the inner one represents the order line.

- DISTINCT is used to avoid that a product appears more than once per order.

- From the given list of products, a random number is chosen where the maximum number of products is restricted to the value of measure [_No of Products per transaction (max)].

Figure 14-5 shows the table Fact - simple and the configuration of the data model.

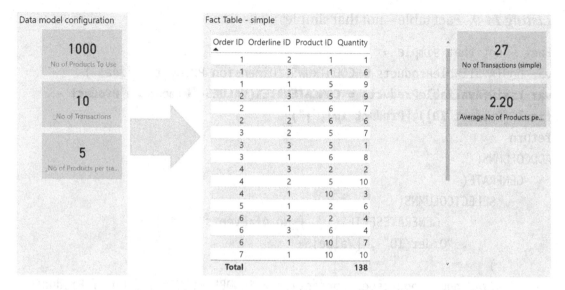

Figure 14-5. *Fact – simple*

Listing 14-8 shows the statement used to configure the sample data model to create the preceding figure.

Listing 14-8. Sample Data Configuration – fact simple

```
Sample Data Configuration =
UNION(
    ROW("Parameter Name" , "No Of Products to use", "Parameter Value" ,
    1000)
    , ROW("Parameter Name" , "No Of Events", "Parameter Value" , 10)
    , ROW("Parameter Name" , "No of Products per event (max)" , "Paramter
    Value" , 5)
)
```

The data model is configured to use all available products (1000), but only five products per event at the maximum. Watching more closely, it can be realized that each event draws from the same bucket of products. This is not what was intended (at least not in this example). This effect is because of the ProductsPath that is defined outside of GENERATE, and next to that, a list is created that always contains the TOPN product IDs. Listing 14-9 is using a different composition of already known components. The important parts are highlighted and will be explained in more detail below the listing.

Listing 14-9. Fact table – not that simple

```
Fact - not that simple =
var noOfAvailableProducts = COUNTROWS('Dimension Product - final')
var ListOfAvaliableProducts = CONCATENATEX(VALUES('Dimension Product -
final'[Product ID]),[Product ID],"|")
return
ADDCOLUMNS(
    GENERATE(
        SELECTCOLUMNS(
                GENERATESERIES(1 , [_No of Events] , 1 )
            , "Order ID" , [Value]
        )
        ,var noofproductspertransaction = RANDBETWEEN(1 , [_No of Products
        per event (max)])
        return
        DISTINCT(
            SELECTCOLUMNS(
                ADDCOLUMNS(
                    GENERATESERIES(1 ,  noofproductspertransaction)

                    , "Product ID" , PATHITEM(ListOfAvaliableProducts ,
                    RANDBETWEEN(1 , PATHLENGTH(ListOfAvaliableProducts)) ,
                    INTEGER)
                )
                , "Orderline ID" , [Value]
                , "Product ID" , [Product ID]
            )
        )
    )
    ,"Quantity" , RANDBETWEEN(1 , 10)
)
```

The line

```
var ListOfAvaliableProducts = CONCATENATEX(VALUES('Dimension Product -
final'[Product ID]),[Product ID],"|")
```

just creates a string that once again can be used by the PATH functions.

The line

```
PATHITEM(ListOfAvaliableProducts , RANDBETWEEN(1 , PATHLENGTH(ListOf
AvaliableProducts)) , INTEGER)
```

randomly draws an item from the string stored inside the variable
ListOfAvaliableProducts, but due to the used separator can be searched by the
PATHITEM function.

Figure 14-6 shows the report using the table Fact – not that simple.

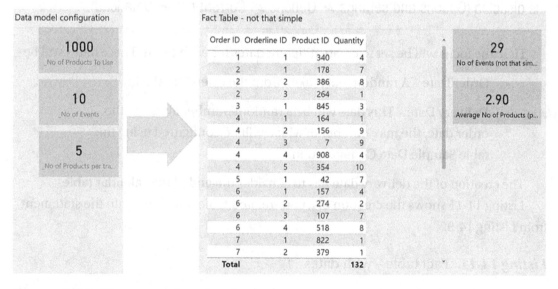

Figure 14-6. *Fact – not that simple*

Different dates

Dates are most essential for almost every analysis. For this reason, we will add a very
simple calendar table to this model.

Listing 14-10 shows how this calendar table is created using DAX.

Listing 14-10. Calendar table basic

```
Calendar table =
    ADDCOLUMNS(
        CALENDAR(
            DATE(2018 , 1 , 1)
            , DATE(2019 , 12 , 31)
        )
        , "Year-Month" , FORMAT([Date] , "YYYYMM")
        , "Year", FORMAT([Date] , "YYYY")
    )
```

Note   The calendar table is marked as a date table, and the option Auto date/time is disabled (Options and settings ➤ Options ➤ Current file ➤ Data load).

Here the focus will be set to create different dates for each event. These dates will be

- Order Date – A random date taken from the calendar table.

- Delivery Date – This date will be a random number ahead of the order date, the max number of which will be configured using the table Sample Data Configuration.

The creation of the delivery date has to consider the end of the calendar table.

Listing 14-11 shows the creation of a date-related table integrated into the statement from Listing 14-9.

Listing 14-11. Fact table – with dates

```
Fact - with Dates =
var noOfAvailableProducts = COUNTROWS('Dimension Product - final')
var ListOfAvaliableProducts = CONCATENATEX(VALUES('Dimension Product -
final'[Product ID]),[Product ID],"|")

var maxDate = MAX('Calendar table'[Date])
var maxDaysAhead = [_No of days ahead]
var maxDateRandom = maxdate - [_No of days ahead]
var c0_base = FILTER(ALL('Calendar table'[Date]) , [Date] <= maxDateRandom)
```

```
var NoOfAvailableDays = COUNTROWS(c0_base)
var c0_base_path = CONCATENATEX(TOPN(NoOfAvailableDays , c0_base, [Date] ,
ASC) , [Date], "|")

var t_OrderID =

                SELECTCOLUMNS(
                    ADDCOLUMNS(
                        GENERATESERIES(1 , [_No of Events] , 1 )
                        , "Order date" , DATEVALUE(PATHITEM(c0_base_path ,
                        RANDBETWEEN(1 , NoOfAvailableDays), TEXT))
                    )
                    , "Order ID" , [Value]
                    , "Order date" , [Order date]
                    , "Delivery date" , [Order date] + RANDBETWEEN(1 ,
                    maxDaysAhead)
                )
return
ADDCOLUMNS(
    GENERATE(
        t_OrderID
        ,var noofproductspertransaction = RANDBETWEEN(1 , [_No of Products
        per events (max)])
        return
        DISTINCT(
            SELECTCOLUMNS(
                ADDCOLUMNS(
                    GENERATESERIES(1 ,  noofproductspertransaction)
                    , "Product ID" , PATHITEM(ListOfAvaliableProducts ,
                    RANDBETWEEN(1 , PATHLENGTH(ListOfAvaliableProducts)) ,
                    INTEGER)
                )
                , "Orderline ID" , [Value]
                , "Product ID" , [Product ID]
            )
        )
```

```
    )
    ,"Quantity" , RANDBETWEEN(1 , 10)
)
```

This statement will be used to draw random days. For each event, an order date and a delivery date will be created. The delivery date depends on the parameter maxDaysAhead and of course on all the dates available. The latest date available is computed:

```
var maxDateRandom = maxdate - [_No of days ahead]
```

This computation makes sure that the delivery date will not exceed the max date of the calendar table.

The available dates will be stored to the table variable c0_base. The available dates will be transformed into an ordered list. This list is stored to the variable c0_base_path.

A final note

The creation of sample data is not only helpful in testing the performance of a measure but also in understanding the problem. There will be times when a measure will not return the expected result.

Keep calm and DAX on.

Index

A

ALL and ALLEXCEPT functions, 113–117

B

Breaking lineage, 65–67

C

CALCULATE function, 331
  context transition, 331–333
  DAX
    calculations, 334
    Studio, 337, 338
  logical plan, 332
  profiler trace, 337
  result of, 331, 332
  running total, 333–337
  SUMX calculation, 331, 332
Calendars
  approaches, 224
  <year_end_date> value
    output, 221
    testing, 222–224
    TOTALYTD function, 220
    YEAR/YTD, 219
Composite *vs.* primitive functions
  CLOSINGBALANCEMONTH, 218
  logical plan, 216
  OPENINGBALANCEMONTH, 218
  TOTALMTD, 216, 217
  types, 215

D

Data modeling, 21
  aspects of, 22
  filter (*see* Filter propagation)
  Power BI project, 21
  relationships, 36–38
    base functions, 32
    cumulative month number, 35
    cumulative value, 34
    edit dialog, 38
    home menu, 37
    matrix value, 35
    one-table solution, 33
    quantity ytd, 32, 35
    relationships.pbix, 37
    table.pbix, 32
    time intelligence working, 33
  single table
    Auto-Exist, 25
    DAX statements, 27
    measurements, 27
    query sent, 29
    report, 26
    rewritten code, 30
    row view, 30
    single table, 26

© Philip Seamark, Thomas Martens 2019
P. Seamark and T. Martens, *Pro DAX with Power BI*, https://doi.org/10.1007/978-1-4842-4897-3

I, J, K

L

M, N, O, P

U

V

W, X, Y, Z

Printed in the United States
By Bookmasters